Cassiane Jrayj De Melo

INTRODUCTION TO REMOTE SENSING IN AGRICULTURE

Cassiane Jrayj De Melo

INTRODUCTION TO REMOTE SENSING IN AGRICULTURE

How technology is revolutionizing the countryside

ScienciaScripts

Imprint

Cover image: www.ingimage.com

This book is a translation from the original published under ISBN 978-3-639-68261-8.

Publisher:
Sciencia Scripts
is a trademark of
Dodo Books Indian Ocean Ltd. and OmniScriptum S.R.L publishing group

120 High Road, East Finchley, London, N2 9ED, United Kingdom
Str. Armeneasca 28/1, office 1, Chisinau MD-2012, Republic of Moldova, Europe
Printed at: see last page
ISBN: 978-620-8-33059-0

INTRODUCTION TO REMOTE SENSING IN AGRICULTURE

CASSIANE JRAYJ DE MELO

2024

ECONOMIC AND ENVIRONMENTAL BENEFITS

Remote sensing offers a wide range of benefits, both economic and environmental. For farmers, one of the main benefits is the ability to increase productivity. By providing accurate data on plant health, soil quality and weather conditions, remote sensing allows farmers to make informed decisions that can lead to higher and more consistent yields.

In addition, technology helps to reduce operating costs. By identifying precisely where and when resources are needed, such as water, fertilisers and pesticides, farmers can avoid waste and apply inputs more efficiently. This not only saves money, but also minimises environmental impact, reducing the amount of chemicals released into the environment.

Sustainability is another area where remote sensing has a significant impact. With the ability to monitor water use and soil management practices, farmers can adopt more sustainable practices that conserve natural resources and protect the environment. In a world where climate change and resource scarcity are growing concerns, adopting technologies that promote sustainability is more important than ever.

SUCCESS STORIES

Throughout this ebook, you'll find several case studies that illustrate how remote sensing is being used successfully in agriculture. From small family farms to large commercial operations, these examples show how the technology can be adapted to meet the specific needs of different types of farmers.

One notable example comes from a wine farm in Rio Grande do Sul. The winery was facing challenges with variability in grape quality due to differences in soil and microclimate in different parts of the vineyard. Using remote sensing, the winery team was able to map these variations and adjust management practices accordingly. This resulted in a more consistent quality grape harvest, improving the overall quality of the wine produced. Another inspiring example comes from an agricultural co-operative in Paraná that has adopted the use of satellites to monitor its corn crops. The cooperative used satellite image data to identify areas with nutrient deficiencies and adjust fertiliser application accordingly. As a result, productivity increased significantly and fertiliser costs were reduced, benefiting all the cooperative's members.

THE JOURNEY STARTS HERE

My intention with this ebook is to provide a comprehensive and accessible introduction to remote sensing in agriculture. Over the next few chapters, we'll explore in detail the various technologies and their practical applications, discuss the benefits and challenges, and share real-life case studies that demonstrate the positive impact of this technology.I want you, the reader, to come away from this reading with a clear understanding of how remote sensing can transform your agricultural operations, whether you are a farmer, student, researcher, consultant or investor. More than that, I want to inspire you to explore and adopt these technologies, taking advantage of the opportunities they offer to increase productivity, reduce costs and promote sustainability.

Remote sensing is a powerful tool that can help tackle many of the challenges facing modern agriculture. From natural resource management to adapting to climate change, this technology offers innovative solutions that can make the difference. I hope this ebook will be a useful and inspiring guide on that journey Let's start exploring the fascinating world of remote sensing in agriculture and discover together how this technology can revolutionise the agricultural sector. Welcome and happy reading!

SUMMARY

and they are often used in combination to provide a complete and detailed view of conditions in the field.

Practical Applications in Agriculture

Remote sensing has numerous practical applications in agriculture, helping farmers to monitor their crops, manage resources and increase productivity. Let's explore some of these applications with real examples.

1. Crop Health Monitoring:
- Using satellite images or drones, farmers can monitor the health of their plants over time. This includes identifying areas of water stress, nutrient deficiency or pest infestation.

- Example: A wheat producer in Paraná uses satellite images to monitor his crops. He can quickly identify areas where the plants are suffering and take corrective measures, such as adjusting irrigation or applying fertilisers.

2. Irrigation Management:
- Thermal and soil moisture sensors help to optimise the use of water, ensuring that the plants receive the right amount without wasting it.

- Example: On a tomato farm in Bahia, soil moisture sensors are used to automatically adjust the irrigation system, saving water and improving plant health.

3. Soil Mapping and Nutrient Management:
- LIDAR and radar sensors are used to map soil topography and identify variations in soil composition. This helps to apply fertilisers and correctives more efficiently.

- Example: A winery in Rio Grande do Sul uses maps of elevation and soil composition to adjust the application of nutrients in its vineyards, resulting in better quality grapes.

4. Harvest and Productivity Forecasting:
- Crop growth models combined with remote sensing data help predict yields and plan harvests.

- Example: A soya cooperative in Mato Grosso uses satellite data to predict the productivity of its crops and plan harvest logistics in advance.

Benefits and Advantages

The benefits of remote sensing in agriculture are numerous. Here are some of the main ones:

1. Increased Productivity:
- By providing accurate, real-time information, remote sensing allows farmers to make informed decisions that lead to higher yields and better quality.

2. Cost reduction:
- The ability to identify exactly where resources are needed, such as water and fertiliser, helps to avoid waste and reduce operating costs.

3. Sustainability:
- More efficient and precise farming practices help conserve natural resources and reduce environmental impact.

4. Data-Based Decision Making:
- The data collected by sensors provides a solid basis for decision-making, replacing intuition and estimation with concrete, accurate information

Challenges and considerations

Despite the many benefits, the implementation of remote sensing in agriculture also presents challenges. Some of the main ones include:

1. Initial cost:
- The initial investment in remote sensing technologies can be high, especially for small farmers. However, costs are decreasing as the technology becomes more accessible.

2. Training and knowledge:
- The effective use of remote sensing requires technical knowledge and training. Farmers and professionals in the sector need to be willing to learn and adopt new technologies.

3. Data integration:
- Collecting data from various sources (satellites, drones, ground sensors) can be complex. Software tools and platforms that integrate this data efficiently are essential.

4. Connectivity and infrastructure:
- In some rural areas, the lack of internet connectivity can make it difficult to transmit and analyse data in real time. Improvements in communication infrastructure are needed to fully capitalise on the benefits of remote sensing.

Conclusion

Remote sensing is a powerful tool that is transforming modern agriculture. From monitoring crop health to precise resource management, the applications are vast and the benefits are significant. By understanding the fundamental concepts and technologies involved, you will be better prepared to exploit and adopt this technology in your own farming operations.

In the following chapters, we'll delve deeper into each of these areas, providing practical examples, case studies and detailed guidance on how to implement remote sensing on your farm or agricultural operation. I am confident that by the end of this ebook you will have a clear and practical understanding of how this technology can help increase productivity, reduce costs and promote sustainability in agriculture.Thank you for joining me on this journey. Let's explore together the fascinating world of remote sensing in agriculture and discover how it can revolutionise the agricultural sector.

CHAPTER 2
THE REMOTE SENSING REVOLUTION IN AGRICULTURE

Welcome to the second chapter of our ebook. Now that you have a basic understanding of what remote sensing is, let's explore how this technology is revolutionising modern agriculture. Remote sensing offers a new way of seeing and managing the land, providing significant economic and environmental benefits.

Impact on Productivity and Efficiency

One of the biggest impacts of remote sensing on agriculture is increased productivity and efficiency. The ability to closely monitor crop health, soil conditions and resource use allows farmers to make more informed and accurate decisions.

For example, imagine a corn producer in the interior of São Paulo. Traditionally, he would need to travel around his property on foot or by tractor to check the condition of the plants, a time-consuming process that is prone to human error. With remote sensing, he can use satellite images or drones to get a detailed view of his entire plantation. The images can reveal areas where plants are suffering from water stress or nutrient deficiency, allowing him to intervene quickly. This results in healthier and more productive crops. Another example comes from a winery in southern Brazil. The winemakers use drones equipped with multispectral cameras to monitor the vineyards. These drones capture images that show the health of the plants and the maturity of the grapes. With this information, the winemakers can adjust irrigation, apply fertilisers precisely and choose the ideal time to harvest. The result is high-quality grapes and superior wines.

Economic Benefits

Remote sensing not only increases productivity, but also offers significant economic benefits. By enabling more precise management of resources, it helps to reduce operating costs and increase profits. Consider a farmer who grows soya in Mato Grosso. Using soil moisture sensors and satellite images, he can monitor the conditions of his land in real time. Instead of applying water evenly across the entire property, he can direct irrigation only to the areas that really need it. This saves water and reduces the energy costs associated with irrigation.

In addition, by identifying problems early, such as pests or diseases, remote sensing allows farmers to intervene before the damage becomes significant. For example, a cotton producer in north-eastern Brazil used drones to detect an initial infestation of caterpillars. With this early detection, he was able to apply pesticides only to the affected areas, saving money and protecting most of the harvest.

Environmental Benefits

Another important aspect of remote sensing is its positive impact on the environment. Modern agriculture faces significant challenges, such as the need to feed a growing population while at the same time preserving natural resources and minimising environmental impacts. environmental impacts. Remote sensing helps to achieve this balance.

One example is sustainable water management. In regions where water is a scarce resource, such as the semi-arid northeast, efficient water use is crucial. Soil moisture sensors and satellite images can help farmers better understand the water needs of their plants and apply the right amount of water at the right time. This not only saves water, but also prevents soil salinisation, a common problem in over-irrigated areas.

In addition, remote sensing helps to reduce the use of agrochemicals. By identifying precisely where pests or diseases are located, farmers can apply pesticides and fungicides in a localised way, instead of spraying the entire plantation. This reduces the amount of chemicals entering the environment, protecting biodiversity and the health of ecosystems.

Use Cases in Various Agricultural Crops

Let's explore some specific use cases for remote sensing in different agricultural crops, to illustrate how this technology can be adapted to meet the specific needs of each type of crop.

1. Coffee:

- Brazil is the largest coffee producer in the world, and remote sensing is helping coffee growers to improve the quality and productivity of their crops. For example, soil sensors and drones are used to monitor soil moisture and temperature, helping to adjust irrigation practices. In addition, multispectral images captured by drones help to identify areas affected by pests or diseases, enabling a quick and effective response

2. Sugarcane:

- Sugarcane is one of Brazil's most important crops, and remote sensing plays a crucial role in its management. Satellite images are used to monitor plant growth and identify areas that need intervention. Soil moisture sensors help optimise irrigation, reducing water consumption. In addition, remote sensing helps to plan the harvest, ensuring that the cane is picked at the ideal time to maximise productivity.

3. Maize and Soya:

- In the Brazilian Cerrado, where large areas are dedicated to growing corn and soya, remote sensing is essential for the efficient management of these crops. Satellite images and drones help monitor plant health, identify nutrient deficiencies and detect pests. Soil sensors provide real-time data on soil moisture and composition, helping to optimise the application of fertilisers and water.

4. Vegetables:

- In vegetable production, where the quality and appearance of products are crucial, remote sensing offers significant benefits. For example, a lettuce producer in São Paulo uses drones to monitor plant health and detect signs of disease early on. This allows him to take preventative measures and maintain the quality of his crops. In addition, humidity sensors help ensure that plants receive the ideal amount of water, avoiding both over- and under-irrigation

The Digital Revolution in Agriculture

The remote sensing revolution in agriculture is part of a larger trend known as digital agriculture or precision agriculture. This trend involves the use of digital technologies to collect, process and analyse data in order to improve decision-making and agricultural management.

A crucial component of digital agriculture is the use of big data and artificial intelligence (AI). The data collected by remote sensing sensors is vast and varied. Analysing this data manually would be impractical, but with the help of AI algorithms, it is possible to process and interpret this information quickly. This allows farmers to receive precise, real-time recommendations on how to improve their farming practices.

For example, on a wheat farm in Rio Grande do Sul, remote sensing data is combined with crop growth models and weather forecasts to optimise harvest planning. AI algorithms analyse this data and provide recommendations on the best time to apply fertiliser, irrigate and harvest, based on specific field conditions and weather forecasts.

Challenges and opportunities

Although remote sensing offers many benefits, its widespread adoption still faces some challenges. One of the main challenges is the initial cost of implementation. Acquiring drones, sensors and specialised software can be expensive, especially for small farmers. However, costs are decreasing as the technology becomes more accessible, and many companies are offering remote sensing solutions as a service, allowing farmers to pay only for use.

Another challenge is the need for training and capacity building. The effective use of remote sensing requires technical knowledge and data interpretation skills. To overcome this challenge, many agricultural organisations are investing in training and education programmes to help farmers adopt these technologies.

Despite these challenges, the opportunities offered by remote sensing are enormous. As technology continues to evolve, we can expect to see even more innovations that will make agriculture more efficient, productive and sustainable. For example, advances in sensors and AI algorithms will enable even more accurate detection of crop problems, and the integration of data from different sources (satellites, drones, soil sensors) will provide an even more complete and detailed view of conditions in the field.

Conclusion

The remote sensing revolution in agriculture has only just begun, but it is already having a profound impact on the sector. By increasing productivity, reducing costs and promoting sustainability, this technology is helping farmers meet the challenges of feeding a growing population and protecting the environment. In this chapter, we explore how remote sensing is transforming modern agriculture, from increased productivity and efficiency to economic and environmental benefits. We saw practical examples of how this technology is being utilised in various agricultural crops and discuss the opportunities and challenges associated with their adoption. In the following chapters, we will continue to deepen our understanding of remote sensing, exploring its specific applications, technologies and implementation methods. I am confident that, by the end of this ebook, you will be well equipped to take advantage of the opportunities that remote sensing will provide.

CHAPTER 3
REMOTE SENSING TECHNOLOGIES

Welcome to the third chapter of our ebook. Here, we'll explore the different remote sensing technologies that are used in agriculture. Each technology has its own advantages and applications, and understanding these differences is essential to making the most of the potential of remote sensing in your agricultural operations.

Introduction to Remote Sensing Technologies

Remote sensing encompasses a variety of technologies, each specialised in collecting different types of data about the earth's surface. The main technologies include optical sensors, thermal sensors, synthetic aperture radar (SAR) and Light Detection and Ranging (LIDAR). Let's explore each one in detail.

Optical sensors

Optical sensors capture images in the range of the electromagnetic spectrum that includes visible light and the near infrared. These sensors are similar to digital cameras, but much more sophisticated and capable of capturing images in several spectral bands.

1. Visible Light and Near Infrared:
- Visible light is the band that our eyes can see, while the near infrared is just beyond red, invisible to the naked eye. Optical sensors capture both bands, allowing the creation of multispectral images.
- Example: In a soya field, optical sensors can detect variations in plant colour, indicating different levels of chlorophyll e, consequently, plant health. Areas with lower near-infrared reflectance may indicate water stress or nutrient deficiency.

2. Vegetation Indices:
- Indices such as NDVI (Normalised Difference Vegetation Index) are calculated from optical images. NDVI is a measure of the quantity and vigour of vegetation, useful for monitoring crop growth.

- Example: A maize grower can use NDVI to identify areas of the field that are less productive and investigate the causes, such as irrigation problems or pest infestation

Thermal sensors

Thermal sensors capture the infrared radiation emitted by the earth's surface, allowing surface temperature to be measured. This is useful for monitoring plant water stress, among other applications.

1. Surface Temperature Measurement:
- Thermal sensors detect the radiation emitted by the surface and calculate the temperature. Plants under water stress tend to heat up more than well-watered plants.

- Example: In a tomato plantation in the north-east of Brazil, thermal sensors are used to monitor the temperature of the plants and identify areas that need additional irrigation, saving water and increasing productivity.

2. Water Stress Detection:
- Changes in plant temperature can indicate water stress. Thermal sensors help detect these changes, enabling a rapid response.

- Example: A winemaker in southern Brazil uses thermal sensors to monitor his vineyards. When specific areas show an increase in temperature, he adjusts the irrigation to ensure that the plants receive the right amount of water

Synthetic Aperture Radar (SAR)

Synthetic aperture radar uses radio waves to create detailed images of the earth's surface. SAR is especially useful because it can penetrate through clouds and work in all weather conditions, day and night.

1. Penetration through clouds and vegetation:
- SAR uses radio waves that can penetrate through clouds, smoke and vegetation, providing data even in adverse weather conditions.

- Example: During the rainy season in the Amazon, optical satellite images can be limited by cloud cover. SAR sensors, however, can provide continuous data on forest health and agricultural expansion.

2. Soil Moisture Measurement:
- SAR sensors can measure soil moisture, crucial information for irrigation management and crop forecasting.
- Example: In a sugar cane plantation in the north-east, SAR sensors help monitor soil moisture, allowing precise adjustments to be made to irrigation and avoiding both excess and lack of water.

Light Detection and Ranging (LIDAR)

LIDAR uses laser pulses to measure distances with high precision, creating detailed maps of the topography and vegetation.

1. Topographical mapping:
- LIDAR emits laser pulses that reflect off the surface and return to the sensor, allowing detailed three-dimensional maps to be created.

- Example: A coffee producer in the mountains of Minas Gerais uses LIDAR to map the topography of his property. With this data, he plans efficient drainage systems and positions his plantations to maximise exposure to the sun.

2. Plant Biomass Detection:
- LIDAR can measure vegetation density and plant height, helping to monitor crop growth and forest health.

- Example: In a forest reserve in Pará, LIDAR is used to monitor forest biomass, helping with conservation and sustainable timber management.

Data Integration and Analysis

One of the greatest advantages of remote sensing is the ability to integrate data from different sources to obtain a complete and detailed view of conditions in the field. Advanced analysis software and digital farming platforms allow farmers to combine data from optical, thermal, SAR and LIDAR sensors to make more informed decisions.

1. Digital Agriculture Platforms:
- Tools such as AgriTech, Climate FieldView and AgriSmart allow data from multiple sources to be integrated and analysed, offering practical, real-time recommendations.

- Example: A wheat producer in Rio Grande do Sul uses a digital agriculture platform to integrate data from satellite images, soil sensors and weather forecasts. With this information, he adjusts his management practices, such as fertiliser application and irrigation, to maximise productivity and minimise environmental impact.

2. Crop Growth Models:
- Integrating remote sensing data with crop growth models helps to predict yields and plan harvests.

- Example: A soya cooperative in Mato Grosso uses remote sensing data to feed crop growth models. This makes it possible to predict harvest yields and plan storage and transport logistics in advance, optimising the entire production chain.

Practical examples of implementation

To illustrate how the different remote sensing technologies are used in practice, let's explore some case studies.

1. Sugarcane cultivation in the Northeast:
- Challenge: Monitoring large plantation areas and optimising irrigation.
- Solution: The company uses a combination of satellite images, thermal sensors and SAR to monitor plant health and soil moisture. Drones equipped with multispectral cameras are used for detailed inspections. The data is integrated into a digital farming platform that provides recommendations in real time.

- Result: The company managed to reduce water use by 20 per cent and increase productivity by 15 per cent, while reducing operating costs.

2. Grape Production in Southern Brazil:
- Challenge: Ensuring the quality of the grapes in variable climatic conditions.

- Solution: Winemakers use drones with optical and thermal sensors to monitor the health of the vines and adjust irrigation and nutrient application. Soil sensors provide real-time data on soil moisture and composition.

- Result: The production of high-quality grapes increased significantly, allowing the creation of award-winning wines and improving the winery's profitability.

3. Soya cultivation in the Cerrado:

- Challenge: Managing soil variability and maximising productivity.

- Solution: LIDAR sensors are used to map topography and identify variations in soil composition. Satellite and SAR data monitor plant health and soil moisture. All the information is integrated into a digital agriculture platform.

- Result: The farmer was able to optimise fertiliser application and irrigation, resulting in a more uniform and productive harvest.

Future of Remote Sensing in Agriculture

As technology continues to evolve, the future of remote sensing in agriculture looks promising. Advances in artificial intelligence, machine learning and big data are enabling more accurate and faster analyses, helping farmers to make even more informed decisions.

1. Artificial Intelligence and Machine Learning:

- AI algorithms are being developed to analyse large volumes of remote sensing data and identify patterns that would be difficult to detect manually

CHAPTER 4
IMPLEMENTING REMOTE SENSING IN AGRICULTURE

Welcome to the fourth chapter of our ebook. Now that we've covered the basics of remote sensing and its technologies, let's explore how to implement these tools in a farming operation. Effective implementation of remote sensing involves planning, selecting the right technologies and integrating data for decision-making.

Implementation Planning

The first step in implementing remote sensing is planning. This process involves defining clear objectives, assessing the specific needs of the agricultural operation and drawing up a detailed plan of action.

1. Definition of Objectives:

- Before starting, it is essential to define the specific objectives of using remote sensing. These objectives can range from increasing productivity, improving irrigation efficiency, detecting pests and diseases early or reducing operating costs.

- Example: A maize producer may set himself the goal of increasing yields by 10 per cent by identifying and correcting nutrient deficiencies in the soil.

2. Needs assessment:

- Assessing the specific needs of the agricultural operation is crucial to selecting the most appropriate technologies. This involves analysing the characteristics of the soil, crops, climate and agricultural practices already in use.

- Example: On a coffee farm in the mountains of Minas Gerais, the uneven topography may require the use of drones to capture detailed aerial images that would not be possible with satellites.

3. Drawing up the Action Plan:

- With the objectives and needs clear, it's time to draw up an action plan. This plan should include a timetable, a definition of responsibilities, the necessary budget and the technological resources to be used.

- Example: An action plan for a soya producer in Mato Grosso might include acquiring drones, hiring data analysis services and training staff to operate the new equipment.

Selecting the Right Technologies

Selecting the right remote sensing technologies is a critical step in ensuring successful implementation. As we saw in the previous chapter, there are several technologies available, each with its own advantages and applications.

1. Satellite images:
- Satellite images are ideal for monitoring large areas and obtaining consistent data over time. They are useful for analysing trends and detecting seasonal changes.

- Example: A wheat producer in Rio Grande do Sul uses satellite images to monitor crop growth throughout the season and plan the harvest accordingly.

2. Drones:
- Drones are extremely versatile and can capture high-resolution images of specific areas. They are particularly useful for detailed inspections and real-time monitoring.

- Example: At a winery in southern Brazil, drones equipped with multispectral cameras are used to monitor the health of the vines and adjust irrigation and nutrient application as necessary.

3. Soil sensors:
- Soil sensors provide accurate data on soil moisture, temperature and composition. They are essential for efficient irrigation management and fertiliser application.

- Example: On a vegetable plantation in São Paulo, soil sensors help ensure that the plants receive the ideal amount of water, avoiding both over- and under-irrigation.

4. Digital Agriculture Platforms:
- Digital agriculture platforms integrate data from various sources and provide analyses and recommendations in real time. They facilitate

informed decision-making and efficient management of the farming operation.

- Example: A sugar cane producer in the north-east of Brazil uses a digital agriculture platform to integrate satellite data, drones and soil sensors, optimising all stages of cultivation.

Data Collection and Integration

Collecting and integrating data are fundamental steps in making the most of remote sensing. The data collected needs to be processed, analysed and integrated into farm management systems to provide valuable insights.

1. Data collection:
- Data collection involves the use of sensors, drones and satellites to capture information about the soil, crops and environment. The frequency of data collection can vary according to the needs of the operation.

- Example: In a soya plantation, drones are used weekly to capture aerial images, while soil sensors provide real-time data on soil moisture and composition.

2. Data Processing:
- The data collected needs to be processed in order to extract useful information. This involves correcting errors, normalising and analysing images and sensor data.

- Example: Satellite images of a cornfield are processed to calculate vegetation indices such as NDVI, which indicate the health of the plants.

3. Data integration:
- Integrating data from different sources provides a holistic view of the agricultural operation. This makes it easier to make data-based decisions and optimise resources.

- Example: On a coffee farm, data from drones, soil sensors and weather forecasts are integrated into a digital agriculture platform that provides recommendations on irrigation, fertiliser application and pest management.

Analysing and Making Decisions

Once the data is collected and integrated, the next step is analysis and decision-making. Advanced analysis tools, such as artificial intelligence and machine learning, can help identify patterns and trends that would not be easily visible otherwise.

1. Data Analysis:
- Data analysis involves the use of algorithms and models to interpret the information collected and provide practical insights.

- Example: In a sugar cane plantation, crop growth models are used to predict productivity and adjust management practices as necessary.

2. Decision-making:
- Based on data analysis, farmers can make informed decisions to improve the productivity, efficiency and sustainability of their operations.

- Example: A vegetable grower uses remote sensing data to adjust irrigation and nutrient application, ensuring that the plants receive proper care and maximising the quality and quantity of the harvest.

Training and Capacity Building

The effective implementation of remote sensing also requires the training and capacity building of staff. It is essential that everyone involved in the agricultural operation understands how to use the technologies and interpret the data.

1. Technical Training:
- Technical training covers the use of equipment, data collection and the operation of analysis software.

- Example: On a soya farm, staff are trained to operate drones and soil sensors, as well as to use digital agriculture platforms to analyse the data collected.

2. Training in Data Analysis:
- Training in data analysis involves developing the skills to interpret information and make decisions based on data.

- Example: A maize producer attends workshops on agricultural data analysis, learning how to use artificial intelligence and machine learning tools to optimise his management practices.

3. Continuous updating:

- Remote sensing technology is constantly evolving, and it's important for farmers to keep up to date with the latest innovations and best practices.

- Example: A winemaker in the south of Brazil regularly attends conferences and seminars on precision agriculture, ensuring that he is always up to date on new technologies and remote sensing methods.

Implementation Case Studies

To illustrate how the implementation of remote sensing can transform agricultural operations, let's explore some case studies.

1. Coffee growing in Minas Gerais:

- Challenge: Monitoring plant health in a mountainous area with difficult access.

- Solution: Drones equipped with multispectral cameras are used to capture detailed aerial images. Soil sensors provide data on soil moisture and composition. All the information is integrated into a digital farming platform.

- Result: The coffee producer was able to increase productivity by 20 per cent and reduce operating costs by 15 per cent, while at the same time improving the quality of the coffee.

2. Soya production in Mato Grosso:

- Challenge: Managing large areas of cultivation and optimising the use of resources.

- Solution: Satellite images are used to monitor crop growth and identify problem areas. Drones are used for detailed inspections and soil sensors provide real-time data on soil moisture and composition.

- Result: The soya grower was able to reduce water use by 25% and increase productivity by 10%, resulting in a more sustainable and profitable operation.

3. Winery in the south of Brazil:

- Challenge: Ensuring the quality of the grapes in variable climatic conditions.
- Solution: Drones with optical and thermal sensors monitor the health of the vines. Soil sensors provide data on the humidity and composition of the soil. All the information is integrated into a digital agriculture platform.
- Result: The quality of the grapes improved significantly, enabling the production of award-winning wines and increasing the winery's profitability.

Conclusion

Implementing remote sensing in agriculture requires planning, selecting the right technologies, collecting and integrating data, analysing and making informed decisions, and continuous staff training. By following these steps, farmers can make the most of remote sensing's potential, improving the productivity, efficiency and sustainability of their operations. Remote sensing technology offers a unique opportunity to transform agriculture, making it more precise, efficient and sustainable. By adopting these technologies, farmers can meet current and future challenges, guaranteeing food security and environmental sustainability for generations to come.

CHAPTER 5
REMOTE SENSING FOR CROP MONITORING

Welcome to the fifth chapter of our ebook. In this chapter, we'll delve into the use of remote sensing for crop monitoring. Continuous crop monitoring is one of the most important and beneficial applications of remote sensing in agriculture, as it allows farmers to keep track of plant development, identify problems early and make informed decisions to optimise productivity.

Importance of Crop Monitoring

Crop monitoring is essential to ensure that plants are growing healthily and efficiently. With remote sensing, farmers can get a comprehensive and detailed view of the state of their crops, which is crucial for:

1. Early detection of problems:
- Identify problems such as nutritional deficiencies, pests, diseases and water stress at an early stage, when it is still possible to intervene and correct the situation.

- Example: In a wheat plantation in Rio Grande do Sul, satellite images revealed areas with lower vegetative vigour. On investigating, the farmer discovered an initial infestation of rust, enabling quick and effective treatment before the disease spread.

2. Informed Decision Making:
- Providing accurate, up-to-date data that helps farmers make informed decisions about irrigation, fertilisation, pest control and other management practices.

- Example: A maize producer in Paraná uses data from optical sensors to monitor plant health and adjust fertiliser application according to the specific needs of each area of the field.

3. Resource optimisation:
- Using resources more efficiently, reducing the waste of water, fertilisers and pesticides, and minimising environmental impact.

- Example: On a coffee farm in Minas Gerais, the use of soil sensors and drones to monitor soil moisture resulted in a significant reduction in water consumption without compromising productivity.

Remote Sensing Tools for Crop Monitoring

There are various remote sensing tools and technologies that can be used for crop monitoring. Each tool offers different types of data and insights, allowing for a complete and detailed view of conditions in the field.

1. Satellite images:

- Satellite images are ideal for monitoring large areas and obtaining consistent data over time. They can be used to create maps vegetation maps, detect seasonal e monitor crop development.

- Example: A soya producer in Mato Grosso uses satellite images to monitor the progress of the crop throughout the season. With this data, he can identify areas with low growth and take corrective action quickly

2. Drones:

- Drones are extremely versatile and can capture high-resolution images of specific areas. They are particularly useful for detailed inspections and real-time monitoring.

- Example: At a winery in southern Brazil, drones equipped with multispectral cameras are used to monitor the health of the vines and adjust irrigation and nutrient application as necessary.

3. Soil Sensors:

- Soil sensors provide accurate data on soil moisture, temperature and composition. They are essential for efficient irrigation management and fertiliser application.

- Example: On a vegetable plantation in São Paulo, soil sensors help ensure that the plants receive the ideal amount of water, avoiding both over- and under-irrigation.

4. Thermal Sensors:

- Thermal sensors capture the infrared radiation emitted by the earth's surface, allowing the surface temperature to be measured. This is useful for monitoring plant water stress.

- Example: In a tomato plantation in the north-east of Brazil, thermal sensors are used to monitor the temperature of the plants and identify areas that need additional irrigation, saving water and increasing productivity.

Vegetation Indices

Vegetation indices are a powerful tool for monitoring crops. They are calculated from remote sensing images and provide valuable information on plant health and vigour.

1. NDVI (Normalised Difference Vegetation Index):
- NDVI is one of the most widely used vegetation indices and is calculated based on the reflectance of the near infrared (NIR) and red light. It ranges from -1 to 1, where higher values indicate greater vegetative vigour.

- Example: In a maize plantation, NDVI is used to identify areas with lower vegetative vigour, which may be suffering from nutritional deficiencies or water stress.

2. EVI (Enhanced Vegetation Index):
- EVI is an improved version of NDVI, which corrects some of the limitations of NDVI in areas with high biomass or exposed soil cover. It is particularly useful in dense forests and agricultural areas with high vegetation.

- Example: A soya producer in the Cerrado uses EVI to monitor plant development in areas with variable ground cover, ensuring a more accurate assessment of crop health.

3. SAVI (Soil Adjusted Vegetation Index):
- SAVI is adjusted to minimise the effect of soil reflectance, which makes it useful in areas with little vegetation or exposed soil. It is calculated in a similar way to NDVI, but includes an adjustment factor for soil.

- Example: In a sugar cane plantation, SAVI is used to monitor plant growth in areas with exposed soil, helping to identify growth problems early on.

Practical Applications of Crop Monitoring

We will explore some practical applications of crop monitoring using remote sensing, illustrating how these technologies can be used to improve agricultural management.

1. Water Stress Detection:
- Water stress can have a significant impact on crop yields. Thermal sensors and vegetation indices such as NDVI are used to identify water-stressed areas and adjust irrigation as necessary.

- Example: In a cotton plantation in north-eastern Brazil, thermal sensors detect areas with water stress, allowing the farmer to adjust irrigation and avoid yield losses.

2. Pest and disease monitoring:
- Remote sensing can help detect pests and diseases early, before they spread and cause significant damage.

High-resolution images and vegetation indices are used to identify affected areas.

- Example: A winemaker in southern Brazil uses drones to capture detailed images of the vines. By analysing the images, he identifies areas with pest infestation and applies specific treatments, preventing spread and minimising the use of pesticides.

3. Nutrient Management:
- Monitoring plant health and soil composition helps optimise fertiliser application. Soil sensors and multispectral images provide data on nutrient deficiencies and enable precise nutrient application.

- Example: In a maize plantation in Paraná, soil sensors detect areas with low nutrient levels. Based on this data, the farmer applies fertiliser in a targeted manner, improving plant health and increasing productivity.

4. Harvest planning:
- Continuous crop monitoring allows for more precise harvest planning. Remote sensing data helps predict crop maturity and plan the harvest at the ideal time.

- Example: A wheat producer in Rio Grande do Sul uses satellite images to monitor plant maturity. With this data, he plans the harvest at the ideal time, guaranteeing maximum quality and yield.

Benefits of Monitoring with Remote Sensing

The use of remote sensing for crop monitoring offers a number of benefits, including:

1. Increased Productivity:
- Continuously monitoring plant health and soil conditions allows problems to be identified early and corrective measures to be taken quickly, resulting in increased productivity.

- Example: A soya producer in Mato Grosso do Sul detects a nutritional deficiency at an early stage and applies specific fertilisers, resulting in a significant increase in productivity.

2. Cost reduction:
- Efficient use of resources such as water, fertilisers and pesticides helps to reduce operating costs. Remote sensing allows precise application of these resources, avoiding waste.

- Example: In a sugar cane plantation in the Northeast, monitoring soil moisture with soil sensors allows for more efficient irrigation, reducing water consumption and associated costs.

3. Environmental sustainability:
- The use of remote sensing technologies contributes to environmental sustainability by enabling more efficient management of natural resources and reducing the environmental impact of agriculture.
- Example: A vegetable grower in São Paulo uses remote sensing data to optimise irrigation and pesticide application, resulting in a more sustainable operation with less environmental impact.

4. Data-Based Decision Making:
- Access to accurate, up-to-date data allows farmers to make informed, evidence-based decisions, improving the management of the agricultural operation.

- Example: A winemaker in southern Brazil uses data from drones and soil sensors to adjust irrigation, nutrient application and pest control, resulting in high-quality wine production.

Conclusion

Remote sensing offers a powerful tool for monitoring crops, allowing farmers to follow plant development, identify problems early on and make informed decisions. Implementing these technologies can result in increased productivity, reduced costs and greater environmental sustainability.

By adopting remote sensing, farmers can transform their operations, making them more efficient and resilient in the face of current and future challenges. With the ability to continuously monitor crops and adjust management practices based on accurate data, remote sensing is becoming an indispensable tool for modern agriculture a.

CHAPTER 6
REMOTE SENSING FOR WATER RESOURCE MANAGEMENT

Welcome to the sixth chapter of our ebook. In this chapter, we'll explore how remote sensing can be used for the efficient management of water resources in agriculture. Water is a vital resource for crop growth, and its proper management is essential to guarantee the productivity and sustainability of agricultural operations.

Importance of Water Resources Management

Efficient water management in agriculture is crucial for several reasons. Firstly, water is a finite resource and, in many regions, it is becoming increasingly scarce due to climate change and increased demand. In addition, excessive irrigation can lead to soil salinisation, environmental degradation and wasted resources. It is therefore essential to use advanced technologies to monitor and manage water efficiently.

How Remote Sensing Helps Water Management

Remote sensing offers various tools and technologies that can be used to monitor soil moisture, predict irrigation needs and ensure efficient water application. Let's explore some of these technologies and their practical applications.

Soil Moisture Sensors

Soil moisture sensors are devices that measure the amount of water present in the soil. These sensors provide accurate, real-time data on soil moisture, allowing for more efficient irrigation management.

- Practical example: On a vegetable plantation in São Paulo, soil moisture sensors are installed at different depths to monitor moisture in various layers of the soil. With this data, the farmer can adjust irrigation according to the specific needs of the plants, avoiding both a shortage and an excess of water.

Satellite Images

Satellite images are a powerful tool for monitoring soil moisture over large areas. Using remote sensing techniques such as near-infrared reflectance (NIR) and thermal radiation, it is possible to obtain detailed information on soil moisture and plant water stress.

- Practical example: On a sugar cane farm in the north-east of Brazil, satellite images are used to monitor soil moisture throughout the property. With this data, the farmer can identify areas that need additional irrigation and optimise water use, resulting in a significant reduction in water consumption.

Drones

Drones are extremely versatile and can be equipped with multispectral and thermal cameras to capture detailed images of soil and plants. They are particularly useful for monitoring small areas or fields with complex topography.

- Practical example: At a winery in southern Brazil, drones are used to monitor soil moisture and the health of the vines. The images captured by the drones help the winemaker to adjust irrigation precisely, ensuring that the plants receive the optimum amount of water.

Irrigation Forecasting Models

Irrigation forecasting models use remote sensing data and other environmental parameters to predict crop irrigation needs. These models take into account factors such as evapotranspiration, rainfall and the soil's water retention capacity to provide precise recommendations on when and how much to irrigate.

- Practical example: In a maize plantation in Paraná, an irrigation forecasting model is used to determine the ideal time to irrigate the plants. Based on the data provided by the model, the farmer can plan irrigation efficiently, saving water and improving productivity.

Benefits of Water Management with Remote Sensing

The use of remote sensing for water resource management offers a number of benefits that contribute to the efficiency, sustainability and productivity of agricultural operations.

Water Saving

One of the main advantages of remote sensing is the ability to save water. By monitoring soil moisture and adjusting irrigation according to the specific needs of the plants, farmers can avoid wasting water and ensure that every drop is used efficiently.

- Practical example: On a rice plantation in Rio Grande do Sul, soil moisture sensors help reduce water consumption by up to 30 per cent, without compromising productivity. This not only saves water, but also reduces the operating costs associated with irrigation.

Increased Productivity

Efficient water management also contributes to increased crop productivity. Well-irrigated plants grow healthier and are less susceptible to stresses such as drought, resulting in larger and better quality harvests.

- Practical example: On a soya farm in Mato Grosso, the use of satellite images to monitor soil moisture results in a 20% increase in productivity, as it allows for more precise and efficient water management.

Environmental sustainability

In addition to the economic benefits, efficient water management also contributes to environmental sustainability. By reducing excessive water use and minimising the environmental impact of irrigation, farmers can operate more sustainably and responsibly.

- Practical example: On a fruit farm in the São Francisco Valley, the implementation of soil moisture sensors and irrigation forecasting models results in a more sustainable operation, with less environmental impact and greater efficiency in the use of water.

Cost reduction

Water savings and increased irrigation efficiency also result in a reduction in operating costs. Less water used means lower costs for pumping, energy and maintenance of irrigation systems. Practical example: On a coffee plantation in Minas Gerais, the use of drones to monitor soil moisture allows for more efficient irrigation. accurate and efficient, resulting in a significant reduction in operating costs.

Challenges and considerations

Although remote sensing offers many advantages for water resource management, there are also challenges that must be considered. Implementing these technologies requires initial investment, training and adaptation to existing management practices.

Cost of Implementation

The purchase and installation of soil moisture sensors, drones and satellite imaging systems can represent a significant cost for farmers. However, the long-term benefits in terms of water savings, increased productivity and reduced costs generally outweigh the initial investment.

Training and Capacity Building

The effective use of remote sensing technologies requires training and capacity building. Farmers need to learn how to interpret the data collected and make informed decisions based on that data.

- Practical example: On a vegetable farm in the interior of São Paulo, a farmer takes part in a training programme on the use of soil moisture sensors and drones. With the knowledge acquired, he is able to implement these technologies effectively, resulting in more efficient water management.

Integration with Existing Management Practices

Integrating remote sensing technologies with existing management practices can be challenging. It is important that farmers adapt their management practices to make the most of the data and insights provided by remote sensing technologies.

- Practical example: On a sugar cane plantation in the north-east, the farmer adjusts his irrigation plan based on data collected by soil moisture sensors and satellite images. This integration results in more efficient irrigation and a more sustainable operation.

Conclusion

The efficient management of water resources is essential for the productivity and sustainability of agriculture. Remote sensing offers a series of tools and technologies that make it possible to monitor soil moisture, predict irrigation needs and optimise water use efficiently.

By adopting these technologies, farmers can save water, increase productivity, reduce costs and operate more sustainably. Although there are challenges in implementing these technologies, the long-term benefits often outweigh the initial costs and efforts.

In this chapter, we explore how remote sensing can be used for water management, highlighting practical examples and tangible benefits. By integrating these technologies into agricultural management practices, farmers can ensure the efficiency and sustainability of their operations, contributing to water and food security in the future.

CHAPTER 7

REMOTE SENSING FOR PEST AND DISEASE MANAGEMENT

Hello and welcome to the seventh chapter of our ebook. In this chapter, we'll explore how remote sensing can be a powerful tool in the management of pests and diseases in agricultural crops. Effective pest and disease control is essential to ensure plant health, crop productivity and the sustainability of agricultural operations.

The Importance of Pest and Disease Management

Pests and diseases are one of the main threats to agriculture and can cause significant losses in crop yield and quality. Detecting and treating these threats early is crucial to minimising damage and ensuring healthy agricultural production.

Impacts of Pests and Diseases

Pests and diseases can have various negative impacts on crops, including:

- Reduced productivity: Infestations can destroy plants or parts of them, resulting in smaller harvests.

- Decreased quality: Fruit and vegetables damaged by pests or diseases may be unsuitable for consumption or sale.

- Increased costs: Chemical or biological control of pests and diseases can be expensive, especially if applied indiscriminately.
- Environmental impact: Excessive use of pesticides can contaminate soil and water, affecting local biodiversity.

How Remote Sensing Helps Pest and Disease Management

Remote sensing enables early detection and continuous monitoring of crop conditions, facilitating efficient pest and disease management. Let's explore some of the most widely used technologies and methodologies.

Satellite Images

Satellite images can be used to monitor large agricultural areas and detect changes in vegetation that may indicate the presence of pests or diseases. This data helps to identify problems early and target specific actions.

- Practical example: In a soya plantation in Mato Grosso, satellite images revealed abnormal reflectance patterns in certain areas, indicating the presence of Asian soya rust. With this information, the farmer was able to apply fungicides in a targeted manner, preventing the spread of the disease and minimising the use of chemical products.

Drones

Drones are extremely effective for detecting and monitoring pests and diseases in specific areas. Equipped with multispectral and thermal cameras, they can capture detailed images of plants, enabling precise analysis of crop conditions.

- Practical example: At a winery in southern Brazil, drones are used to monitor the health of the vines. Multispectral images revealed areas with symptoms of mildew, allowing quick and specific treatment, preserving the health of the vines and the quality of the grapes

Thermal sensors

Thermal sensors capture the infrared radiation emitted by plants, making it possible to detect water stress and diseases. Stressed plants generally emit more heat, which can be detected by these sensors.

- Practical example: In a tomato plantation in the north-east of Brazil, thermal sensors are used to monitor the temperature of the plants. The detection of areas with abnormally high temperatures indicated the presence of a spider mite infestation, enabling rapid and effective intervention.

Pest and Disease Forecasting Models

Forecasting models use remote sensing data and other environmental parameters to predict the occurrence of pests and diseases. These models help farmers take preventative measures before problems become serious.

- Practical example: In a wheat plantation in Rio Grande do Sul, a disease forecast model based on soil moisture, temperature and satellite image data indicated a high risk of powdery mildew. The farmer applied fungicides preventatively, avoiding a significant infestation and protecting the crop.

Benefits of Remote Sensing in Pest and Disease Management

The use of remote sensing for pest and disease management offers several benefits, including:

Early Detection

The ability to detect pests and diseases at an early stage is one of the greatest benefits of remote sensing. This allows farmers to take corrective action before problems spread and cause significant damage.

- Practical example: In a coffee plantation in Minas Gerais, the early detection of coffee berry borers using drone images enabled specific and effective treatment, resulting in lower production losses and a reduction in the use of pesticides.

Precise Application of Treatments

Remote sensing enables more precise application of pesticides and fungicides, reducing the use of chemicals and the associated costs, as well as minimising environmental impact.

- Practical example: On a vegetable farm in the interior of São Paulo, drones are used to apply pesticides only in areas affected by pests, resulting in a significant reduction in the use of chemicals and less environmental impact.

Continuous Monitoring

Continuous crop monitoring allows for constant vigilance, helping to identify and treat problems as they arise. This is especially important in high-value crops, where losses can be significant.

- Practical example: On an apple plantation in southern Brazil, the use of thermal sensors and drones to monitor tree health has resulted in more effective management of pests such as fruit flies, improving the quality and quantity of the harvest.

Cost savings

Early detection and the precise application of treatments result in operational cost savings, reducing the need for extensive and expensive interventions.

- Practical example: On a cotton plantation in north-eastern Brazil, the use of pest forecasting models and satellite images to monitor plant health resulted in significant savings in pest control costs, increasing the profitability of the operation.

Challenges and considerations

Although remote sensing offers many benefits, there are also challenges and considerations that must be taken into account when implementing it.

Cost of Implementation

Acquiring and implementing remote sensing technologies can represent a significant cost, especially for small farmers. However, the long-term benefits in terms of increased productivity and reduced costs usually outweigh the initial investment.

Training needs

The effective use of remote sensing technologies requires training and capacity building. Farmers need to learn how to interpret the data collected and make informed decisions based on that data.

- Practical example: On a soya plantation in Mato Grosso, the farmer took part in a training programme on the use of drones and satellite images. With the knowledge he acquired, he was able to implement these technologies effectively, resulting in more efficient pest and disease management.

Integration with Existing Management Practices

Integrating remote sensing technologies with existing management practices can be challenging. It is important that farmers adapt their management practices to make the most of the data and insights provided by remote sensing technologies.

- Practical example: On a grape plantation in southern Brazil, the winegrower adjusted his management practices based on data collected by drones and thermal sensors. This integration resulted in more efficient pest and disease management, improving grape quality and wine production.

Conclusion

Remote sensing offers a powerful tool for pest and disease management in agriculture, enabling early detection, continuous monitoring and precise application of treatments. Implementing these technologies can result in increased productivity, reduced costs and greater environmental sustainability. By adopting remote sensing, farmers can transform their operations, making them more efficient and resilient in the face of pest and disease challenges. With the ability to continuously monitor crops and adjust management practices based on accurate data, remote sensing has become an indispensable tool for modern agriculture. In this chapter, we explore how remote sensing can be used for pest and disease management, highlighting practical examples and tangible benefits. By integrating these technologies into agricultural management practices, farmers can ensure the health and productivity of their crops, contributing to food security and environmental sustainability in the future.

CHAPTER 8
REMOTE SENSING FOR SOIL MONITORING

Welcome to the eighth chapter of our ebook. In this chapter, we'll explore how remote sensing can be used for soil monitoring in agriculture. Soil health is fundamental to crop productivity and the sustainability of agricultural operations. Using advanced remote sensing technologies, farmers can obtain detailed information about soil conditions, enabling more efficient and sustainable management.

Importance of Soil Monitoring

Soil is one of the most valuable resources in agriculture, as it provides essential nutrients, water and physical support for plant growth. Monitoring soil health and conditions is crucial to ensuring crop productivity and the sustainability of agricultural operations.

Important Soil Parameters

Several soil parameters are important for healthy plant growth, including:

- Soil texture: Determines the soil's ability to retain water and nutrients.

- Soil moisture: Influences the availability of water for plants.
- Soil pH: Affects nutrient availability and plant health.
- Nutrient levels: Essential for plant growth, including nitrogen, phosphorus and potassium.

- Organic matter: Improves soil structure and water and nutrient retention.

Remote Sensing Technologies for Soil Monitoring

Remote sensing offers various technologies that can be used to monitor soil conditions, providing accurate, real-time data for farmers. Let's explore some of these technologies and their practical applications.

Sun humidity sensors

Soil moisture sensors are devices that measure the amount of water present in the soil. These sensors provide accurate, real-time data on soil moisture, allowing for more efficient irrigation management.

- Practical example: In a maize plantation in Paraná, soil moisture sensors are installed at different depths to monitor the moisture in various layers of the soil. With this data, the farmer can adjust the irrigation according to the specific needs of the plants, avoiding both a shortage and an excess of water.

Satellite Images

Satellite images are a powerful tool for monitoring soil health over large areas. Using remote sensing techniques such as near-infrared reflectance (NIR) and thermal radiation, it is possible to obtain detailed information about the texture, moisture and health of the soil.

- Practical example: On a soya farm in Mato Grosso, satellite images are used to monitor soil moisture throughout the property. With this data, the farmer can identify areas that need additional irrigation and optimise water use, resulting in a significant reduction in water consumption.

Drones

Drones are extremely versatile and can be equipped with multispectral and thermal cameras to capture detailed images of soil and plants. They are particularly useful for monitoring small areas or fields with complex topography.

- Practical example: At a winery in southern Brazil, drones are used to monitor soil moisture and the health of the vines. The images captured by the drones help the winemaker to adjust irrigation precisely, ensuring that the plants receive the optimum amount of water.

Electrical Conductivity Sensors

Electrical conductivity sensors measure the soil's ability to conduct electricity, which is related to soil texture, salinity, moisture and nutrient levels. These sensors provide valuable data on soil conditions, allowing for more efficient management.

- Practical example: In a cotton plantation in north-eastern Brazil, electrical conductivity sensors are used to map soil variability. With this data, the farmer can adjust fertiliser application and irrigation according to the specific needs of each area, resulting in greater efficiency and productivity.

Laboratory Analyses

Although not a form of remote sensing, laboratory soil analyses complement the data obtained by sensors and images, providing detailed information on the chemical composition and fertility of the soil. Soil samples are collected and analysed in the laboratory to determine nutrient levels, pH and organic matter.

- Practical example: On a vegetable farm in the interior of São Paulo, soil samples are collected periodically and sent for laboratory analysis. The results are combined with data from soil moisture sensors and satellite images to provide a comprehensive view of soil health, allowing for more precise and efficient management.

Benefits of Remote Sensing in Monitoring the Sun

The use of remote sensing for soil monitoring offers several benefits that contribute to the efficiency, sustainability and productivity of agricultural operations.

Increased Productivity

Continuous and accurate monitoring of soil conditions allows farmers to adjust their management practices according to the specific needs of the crops, resulting in increased productivity.

- Practical example: In a wheat plantation in Rio Grande do Sul, the use of soil moisture sensors and satellite images resulted in a 15 per cent increase in productivity, as it allowed for more precise and efficient water management.

Saving Resources

The efficient use of water and fertilisers is crucial to the sustainability of agricultural operations. Remote sensing allows farmers to apply these resources precisely, avoiding waste and reducing operating costs

- Practical example: On a sugar cane farm in the north-east, the use of soil moisture and electrical conductivity sensors has resulted in significant savings in water and fertiliser, improving efficiency and reducing costs.

Environmental sustainability

Efficient management of soil resources contributes to environmental sustainability by reducing the negative impact of agriculture on the environment. The use of remote sensing allows farmers to practise more sustainable and responsible agriculture.

- Practical example: On a fruit plantation in the São Francisco Valley, the implementation of soil moisture and electrical conductivity sensors has resulted in a more sustainable operation, with less environmental impact and greater efficiency in the use of resources.

Informed decision-making

Access to accurate, real-time data on soil conditions allows farmers to make informed, evidence-based decisions, improving the management of the agricultural operation.

- Practical example: On a vegetable farm in the interior of São Paulo, the farmer uses data from soil moisture sensors and images of satellite to adjust irrigation, nutrient application and crop management, resulting in a more efficient and productive operation .

Challenges and considerations

Although remote sensing offers many benefits, there are also challenges and considerations that must be taken into account when implementing it.

Cost of Implementation

Acquiring and implementing remote sensing technologies can represent a significant cost, especially for small farmers. However, the long-term benefits in terms of increased productivity and reduced costs usually outweigh the initial investment.

- Practical example: On a maize plantation in Paraná, the farmer invested in soil moisture sensors and satellite images. Although the initial cost was significant, the benefits in terms of water savings and increased productivity resulted in a return on investment within a few years.

Training needs

The effective use of remote sensing technologies requires training and capacity building. Farmers need to learn how to interpret the data collected and make informed decisions based on that data.

- Practical example: On a soya farm in Mato Grosso, the farmer took part in a training programme on the use of soil moisture sensors and satellite images. With the knowledge acquired, he was able to implement these technologies effectively, resulting in more efficient soil management.

Integration with Existing Management Practices

Integrating remote sensing technologies with existing management practices can be challenging. It is important that farmers adapt their management practices to make the most of the data and insights provided by remote sensing technologies.

- Practical example: On a grape plantation in southern Brazil, the grower adjusted his management practices based on data collected by soil moisture sensors and satellite images. This integration resulted in more efficient soil management, improving plant health and productivity.

Conclusion

Soil monitoring is essential for agricultural productivity and sustainability. Remote sensing offers a series of tools and technologies that make it possible to monitor soil conditions accurately and in real time. By adopting these technologies, farmers can increase productivity, save resources, operate more sustainably and make informed decisions based on data. Although there are challenges in implementing these technologies, the long-term benefits often outweigh the initial costs and obstacles. In this chapter, we explore how remote sensing can be used for soil monitoring, highlighting practical examples and tangible benefits. By integrating these technologies into agricultural management practices, farmers can guarantee the health and productivity of their crops, contributing to food security and environmental sustainability in the future.

CHAPTER 9
REMOTE SENSING IN WATER MANAGEMENT

Hello and welcome to the ninth chapter of our ebook. Here, we're going to delve into the fascinating world of remote sensing applied to water management in agriculture. Water is a vital resource for plant growth and the sustainability of agricultural operations. However, water scarcity and inefficient use can jeopardise crop productivity and the health of the environment. Using advanced remote sensing technologies, farmers can manage water more efficiently, ensuring the sustainability and productivity of their operations.

The Importance of Water Management in Agriculture

Efficient water management is essential for agriculture, as it directly affects plant health and crop productivity. Inadequate irrigation, whether due to excess or lack of water, can result in significant losses in crop yield and quality.

Impacts of Inefficient Irrigation

Inefficient irrigation can have several negative impacts, including:
- Water stress: Lack of water can cause stress in plants, resulting in reduced growth and lower productivity.

- Excess water: Excess irrigation can lead to waterlogged soil, damaging root health and increasing vulnerability to disease.

- Waste of resources: Excessive application of water increases operating costs and wastes a precious resource.
- Environmental impact: Improper use of water can lead to soil and groundwater contamination, affecting the local ecosystem.

Remote Sensing Technologies for Water Management

Remote sensing offers various technologies that can be used to monitor and manage water use in agriculture, providing accurate, real-time data for farmers. Let's explore some of these technologies and their practical applications.

Satellite Images

Satellite images are a powerful tool for monitoring soil moisture and plant health over large areas. Using remote sensing techniques such as near-infrared reflectance (NIR) and thermal radiation, it is possible to obtain detailed information on soil moisture and plant water requirements.

- Practical example: On a rice farm in southern Brazil, satellite images are used to monitor soil moisture and plant health. With this data, the farmer can adjust irrigation according to the specific needs of the plants, avoiding both a shortage and an excess of water.

Drones

Drones are extremely versatile and can be equipped with multispectral and thermal cameras to capture detailed images of soil and plants. They are particularly useful for monitoring small areas or fields with complex topography.

- Practical example: On a grape plantation in the São Francisco Valley, drones are used to monitor soil moisture and the health of the vines. The images captured by the drones help the viticulturist to adjust the irrigation precisely, ensuring that the plants receive the ideal amount of water.

Soil Moisture Sensors

Soil moisture sensors are devices that measure the amount of water present in the soil. These sensors provide accurate, real-time data on soil moisture, allowing for more efficient irrigation management.

- Practical example: In a maize plantation in Paraná, soil moisture sensors are installed at different depths to monitor the moisture in various layers of the soil. With this data, the farmer can adjust the irrigation according to the specific needs of the plants, avoiding both a shortage and an excess of water.

Water Forecasting Models

Forecasting models use remote sensing data and other environmental parameters to predict crop water requirements. These models help farmers plan and manage irrigation more efficiently.

- Practical example: On a wheat plantation in Rio Grande do Sul, a water forecasting model based on soil moisture data, temperature and satellite images indicated an increase in the plants' water needs. The farmer adjusted the irrigation according to the forecasts, ensuring that the plants received the optimum amount of water.

Benefits of Remote Sensing in Water Management

The use of remote sensing for water management offers several benefits that contribute to the efficiency, sustainability and productivity of agricultural operations

Increased Productivity

Continuous and accurate monitoring of water conditions allows farmers to adjust their irrigation practices according to the specific needs of the crops, resulting in increased productivity.

- Practical example: In a soya plantation in Mato Grosso, the use of soil moisture sensors and satellite images resulted in a 20% increase in productivity, as it allowed for more precise and efficient water management.

Saving Resources

The efficient use of water is crucial to the sustainability of agricultural operations. Remote sensing allows farmers to apply water precisely, avoiding waste and reducing operating costs.

- Practical example: On a sugar cane farm in the Northeast, the use of soil moisture sensors and drones resulted in significant water savings, improving efficiency and reducing costs.

Environmental sustainability

Efficient water management contributes to environmental sustainability by reducing the negative impact of agriculture on the environment. The use remote sensing allows farmers to practise more sustainable and responsible agriculture.

- Practical example: On a fruit plantation in the São Francisco Valley, the implementation of soil moisture sensors and drones has resulted in a more sustainable operation, with less environmental impact and greater efficiency in water use

Informed decision-making

Access to accurate, real-time data on water conditions allows farmers to make informed, evidence-based decisions, improving the management of the agricultural operation.

- Practical example: On a vegetable farm in the interior of São Paulo, the farmer uses data from soil moisture sensors and satellite images to adjust irrigation, ensuring that the plants receive the optimum amount of water and avoiding both under- and over-watering.

Challenges and considerations

Although remote sensing offers many benefits, there are also challenges and considerations that must be taken into account when implementing it.

Cost of Implementation

Acquiring and implementing remote sensing technologies can represent a significant cost, especially for small farmers. However, the long-term benefits in terms of increased productivity and reduced costs usually outweigh the initial investment.

- Practical example: On a maize plantation in Paraná, the farmer invested in soil moisture sensors and drones. Although the initial cost was significant, the benefits in terms of water savings and increased productivity resulted in a return on investment in just a few years.

Training needs

The effective use of remote sensing technologies requires training and capacity building. Farmers need to learn how to interpret the data collected and make informed decisions based on that data.

- Practical example: On a soya farm in Mato Grosso, the farmer took part in a training programme on the use of soil moisture sensors and drones. With the knowledge he acquired, he was able to implement these technologies effectively, resulting in more efficient water management.

Integration with Existing Management Practices

Integrating remote sensing technologies with existing management practices can be challenging. It is important that farmers adapt their management practices to make the most of the data and insights provided by remote sensing technologies.

- Practical example: On a grape plantation in southern Brazil, the grower adjusted his management practices based on data collected by soil moisture sensors and drones. This integration resulted in more efficient water management, improving plant health and productivity.

Conclusion

Efficient water management is essential for agricultural productivity and sustainability. Remote sensing offers a series of tools and technologies that make it possible to monitor and manage water use precisely and in real time.

By adopting these technologies, farmers can increase productivity, save resources, operate more sustainably and make informed decisions based on data. Although there are challenges in implementing these technologies, the long-term benefits often outweigh the initial costs and obstacles.

In this chapter, we explore how remote sensing can be used for water management, highlighting practical examples and tangible benefits. By integrating these technologies into agricultural management practices, farmers can ensure the health and productivity of their crops, contributing to food security and environmental sustainability in the future.

With remote sensing, agriculture can take a significant step towards a more efficient, productive and sustainable future, where water, such a precious resource, is used in the most efficient way possible.

CHAPTER 10

REMOTE SENSING FOR DETECTING PESTS AND DISEASES

Welcome to the tenth chapter of our ebook. In this chapter, we'll explore how remote sensing can be a powerful tool in the early detection of pests and diseases in agricultural crops. Pests and diseases can cause significant damage to crops, resulting in yield and quality losses. By utilising remote sensing technologies, farmers can quickly identify problems and take corrective action before losses become severe.

The Importance of Early Detection

Early detection of pests and diseases is crucial for the health of crops and the efficiency of agricultural production. Rapid intervention can prevent the spread of problems and minimise the economic impact.

Impacts of Pests and Diseases

Pests and diseases can have various negative impacts, including:

- Reduced productivity: Pests and diseases can damage plants, reducing crop yields.

- Loss of quality: Infestation can affect the quality of agricultural products, resulting in lower market value.

- Increased costs: Pest and disease control can be expensive, increasing production costs.
- Environmental impact: Excessive use of pesticides and fungicides can have negative effects on the environment and human health.

Remote Sensing Technologies for Pest and Disease Detection

Remote sensing offers various technologies that can be used for the early detection of pests and diseases in agricultural crops. Let's explore some of these technologies and their practical applications.

Satellite Images

Satellite images are a valuable tool for monitoring large areas of cultivation. Using specific spectral bands, such as the near infrared (NIR) and mid infrared (MIR), it is possible to detect changes in plant health that may indicate the presence of pests or diseases.

- Practical example: In a soya plantation in Mato Grosso, satellite images are used to monitor plant health. Changes in spectral reflectance indicated the presence of a caterpillar infestation, allowing the farmer to take corrective measures before the infestation spreads.

Drones

Drones equipped with multispectral and thermal cameras can provide detailed images of the state of crops, enabling early detection of pests and diseases in specific areas of the field.

- Practical example: In a vineyard in the São Francisco Valley, drones are used to monitor the health of the vines. Thermal images captured by the drones revealed water-stressed areas that were more susceptible to pest attacks. With this information, the winegrower was able to apply targeted treatments, preventing further damage.

Sensors in the Field

Sensors installed in the field can continuously monitor the environment and the plants, detecting signs of stress that may indicate the presence of pests or diseases. These sensors can measure parameters such as temperature, humidity and chlorophyll levels in plants.

- Practical example: On a tomato plantation in the interior of São Paulo, chlorophyll sensors are used to monitor plant health. A decrease in chlorophyll levels indicated a fungal infection at an early stage, allowing the farmer to apply fungicides before the disease spread.

Benefits of Remote Sensing in Pest and Disease Detection

The use of remote sensing to detect pests and diseases offers several benefits that contribute to the efficiency, sustainability and productivity of agricultural operations.

Early Detection and Rapid Intervention

Continuous and accurate monitoring allows farmers to detect problems at an early stage and take corrective action quickly, minimising damage to crops.

- Practical example: On a maize farm in Paraná, the use of drones equipped with multispectral cameras enabled the early detection of an infestation of aphids. The farmer applied specific insecticides, preventing a significant loss of yield.

Reducing Pesticide Use

With early detection and targeted application of treatments, farmers can reduce the use of pesticides, cutting costs and environmental impact.

- Practical example: On a coffee plantation in the south of Minas Gerais, field sensors detected the presence of the coffee berry borer at an early stage. The farmer applied treatments only to the affected areas, reducing the use of pesticides and preserving the environment.

Increased Productivity and Quality

Quick and effective intervention in the presence of pests and diseases ensures that plants remain healthy, resulting in higher yields and crop quality.

- Practical example: On a sugar cane farm in the north-east, the use of satellite images enabled early detection of orange rust, a fungal disease. The farmer applied specific fungicides, preserving the health of the plants and guaranteeing a bumper harvest.

Informed decision-making

Access to accurate, real-time data allows farmers to make informed, evidence-based decisions, improving the management of the agricultural operation.

- Practical example: On a vegetable plantation in the interior of the state of Goiás, the farmer uses data from field sensors and drones to monitor the health of the plants. With this information, he can adjust his management practices precisely, ensuring that the plants remain healthy and productive.

Challenges and considerations

Although remote sensing offers many benefits, there are also challenges and considerations that must be taken into account when implementing it.

Cost of Implementation

Acquiring and implementing remote sensing technologies can represent a significant cost, especially for small farmers. However, the long-term benefits in terms of increased productivity and reduced costs usually outweigh the initial investment.

- Practical example: On a wheat plantation in Rio Grande do Sul, the farmer invested in drones and field sensors. Although the initial cost was high, the benefits in terms of early detection of pests and diseases resulted in a return on investment within a few years.

Training needs

The effective use of remote sensing technologies requires training and capacity building. Farmers need to learn how to interpret the data collected and make informed decisions based on that data.

- Practical example: On a soya farm in Mato Grosso, the farmer took part in a training programme on the use of drones and field sensors. With the knowledge he has acquired, he has been able to implement these technologies effectively, resulting in more efficient pest and disease management.

Integration with Existing Management Practices

Integrating remote sensing technologies with existing management practices can be challenging. It is important that farmers adapt their management practices to make the most of the data and insights provided by remote sensing technologies.

- Practical example: On a grape plantation in southern Brazil, the grower adjusted his management practices based on data collected by drones and field sensors. This integration resulted in more efficient pest and disease management, improving plant health and productivity.

Conclusion

Early detection of pests and diseases is essential for agricultural productivity and sustainability. Remote sensing offers a series of tools and technologies that make it possible to monitor and identify problems accurately and in real time. By adopting these technologies, farmers can increase productivity, save resources, operate more sustainably and make informed decisions based on data. Although there are challenges in implementing these technologies, the long-term benefits often outweigh the initial costs and obstacles. In this chapter, we explore how remote sensing can be used to detect pests and diseases, highlighting practical examples and tangible benefits. By integrating these technologies into agricultural management practices, farmers can ensure the health and productivity of their crops, contributing to food security and environmental sustainability in the future. With remote sensing, agriculture can take a significant step towards a more efficient, productive and sustainable future, where pests and diseases are detected and controlled effectively and precisely.

CHAPTER 11

REMOTE SENSING IN WEATHER FORECASTING AND ITS IMPACT ON AGRICULTURE

Welcome to the eleventh chapter of our ebook. In this chapter, we'll explore how remote sensing can be used for weather forecasting and how this directly impacts agriculture. Weather conditions are one of the most critical factors affecting agricultural production. Understanding and predicting the weather can help farmers better plan their activities, reduce risks and increase productivity.

The Importance of Weather Forecasting in Agriculture

Weather forecasting is essential for agriculture, as it allows farmers to make informed decisions about planting, irrigation, harvesting and other management practices. Adverse weather conditions such as drought, heavy rain and frost can cause significant crop losses.

Impacts of Adverse Weather Conditions

Adverse weather conditions can have a number of negative impacts, including:

- Reduced productivity: Heavy rain, drought or frost can damage plants and reduce crop yields.

- Loss of quality: Extreme weather conditions can affect the quality of agricultural products, resulting in lower market value.

- Increased costs: Adverse weather events can require expensive mitigation measures, such as additional irrigation or frost protection.

- Environmental impact: Inadequate adaptation to climatic conditions can lead to unsustainable agricultural practices and environmental degradation.

Remote Sensing Technologies for Weather Forecasting

Remote sensing offers various technologies that can be used to monitor and predict weather conditions. Let's explore some of these technologies and their practical applications.

Meteorological Satellites

Meteorological satellites are powerful tools for monitoring atmospheric conditions on a global scale. They collect data on temperature, humidity, atmospheric pressure, winds and cloud cover, which are used to forecast the weather.

- Practical example: On a corn farm in Mato Grosso, data from meteorological satellites is used to predict heavy rainfall. With this information, the farmer can plan the harvest before the storm arrives, avoiding productivity losses.

Automatic Weather Stations

Automatic weather stations are installed in the field and provide accurate, real-time data on local weather conditions, including temperature, humidity, rainfall and wind speed.

- Practical example: On a coffee plantation in the south of Minas Gerais, an automatic weather station continuously monitors climatic conditions. The data collected helps the farmer plan irrigation and frost protection, guaranteeing the health of the plants.

Climate Models

Climate models use remote sensing data and other environmental parameters to predict future climate conditions. These models can provide short-, medium- and long-term forecasts, helping farmers to plan their activities.

- Practical example: On a wheat plantation in Rio Grande do Sul, a climate model predicted a prolonged drought. With this information, the farmer decided to plant more drought-resistant varieties of wheat and implemented an efficient irrigation system, mitigating the negative impacts of the drought.

Drones

Drones equipped with meteorological sensors can provide detailed data on weather conditions in specific areas of the field. They are particularly useful for monitoring microclimates and local conditions that may not be captured by satellites or weather stations.

- Practical example: On a vegetable plantation in the interior of São Paulo, drones are used to monitor the temperature and humidity of the air in different parts of the field. The data collected helps the farmer adjust irrigation and frost protection, ensuring that the plants receive proper care.

Benefits of Remote Sensing in Weather Forecasting

The use of remote sensing for weather forecasting offers several benefits that contribute to the efficiency, sustainability and productivity of agricultural operations.

Planning Ahead

Access to accurate weather forecasts allows farmers to plan their activities in advance, minimising risks and maximising efficiency.

- Practical example: On a soya farm in Mato Grosso, the weather forecast indicated the arrival of a cold front. The farmer adjusted the planting schedule to avoid exposing the plants to low temperatures, ensuring healthy growth.

Risk Reduction

Weather forecasting allows farmers to identify and mitigate climate risks, such as droughts, heavy rains and frosts, before they cause significant damage.

- Practical example: On a grape plantation in the São Francisco Valley, the weather forecast indicated the possibility of frost. The grower implemented protective measures, such as covering the vines and using heaters, to prevent damage to the plants.

Increased Productivity

Informed decision-making based on climate forecasts allows farmers to optimise their management practices, resulting in higher crop yields and quality.

- Practical example: On a sugar cane farm in the north-east, the use of climate forecasts enabled the irrigation schedule to be optimised. With the precise application of water, the plants grew healthily, resulting in an abundant harvest.

Environmental sustainability

The efficient use of resources and adaptation to climatic conditions contribute to environmental sustainability, reducing the negative impact of agriculture on the environment.

- Practical example: On a fruit plantation in the south of Brazil, the use of climate forecasts enabled the optimisation of water and fertiliser use. With more sustainable management practices, the farmer has reduced environmental impact and improved soil health.

Challenges and considerations

Although remote sensing offers many benefits, there are also challenges and considerations that must be taken into account when implementing it.

Forecast accuracy

The accuracy of climate forecasts can vary depending on the quality of the data and the models used. It is important for farmers to take this variability into account when making decisions based on climate forecasts.

- Practical example: On a maize plantation in Paraná, the weather forecast indicated heavy rainfall, but the rains didn't come as predicted. The farmer adjusted his management practices to deal with the discrepancy, ensuring that the plants didn't suffer from the lack of water.

Cost of Implementation

The acquisition and implementation of remote sensing technologies can represent a significant cost, especially for small companies. farmers. However, the long-term benefits in terms of increased productivity and reduced costs usually outweigh the initial investment.

- Practical example: On a wheat plantation in Rio Grande do Sul, the farmer invested in an automatic weather station and a climate model. Although the initial cost was high, the benefits in terms of planning and risk mitigation resulted in a return on investment in just a few years.

Training needs

The effective use of remote sensing technologies requires training and capacity building. Farmers need to learn how to interpret the data collected and make informed decisions based on that data.

- Practical example: On a soya farm in Mato Grosso, a farmer took part in a training programme on the use of weather satellites and climate models. With the knowledge he acquired, he was able to implement these technologies effectively, resulting in more efficient management of agricultural practices.

Integration with Existing Management Practices

Integrating remote sensing technologies with existing management practices can be challenging. It is important that farmers adapt their management practices to make the most of the data and insights provided by remote sensing technologies.

- Practical example: On a grape plantation in the south of Brazil, the grower has adjusted his management practices based on the weather forecasts. This integration has resulted in more efficient plant management, improving vine health and productivity.

Conclusion

Weather forecasting is essential for agricultural productivity and sustainability. Remote sensing offers a series of tools and technologies that make it possible to monitor and predict weather conditions accurately and in real time.

By adopting these technologies, farmers can increase productivity, save resources, operate more sustainably and make informed decisions based on data. Although there are challenges in implementing these technologies, the long-term benefits often outweigh the initial costs and obstacles.

In this chapter, we explore how remote sensing can be used for weather forecasting, highlighting practical examples and tangible benefits. By integrating these technologies into agricultural management practices, farmers can ensure the health and productivity of their crops, contributing to food security and environmental sustainability in the future.

With remote sensing, agriculture can take a significant step towards a more efficient, productive and sustainable future, where weather conditions are monitored and predicted effectively and accurately.

CHAPTER 12
REMOTE SENSING IN WATER RESOURCES MANAGEMENT

Hello again! In the previous chapter, we discussed weather forecasting and its impact on agriculture. Now, let's explore an equally crucial topic: managing water resources using remote sensing. Water is a vital resource for agriculture, and its efficient management can make the difference between a bountiful harvest and significant losses.

The Importance of Water Management in Agriculture

Water is essential for plant growth, directly influencing agricultural productivity. With climate change and the growing demand for food, efficient water management has become a priority. Using water sustainably helps preserve this vital resource for future generations and increases the resilience of farms to adverse climatic conditions.

Challenges in Water Resources Management

Water resource management faces a number of challenges:

- Water scarcity: In many regions, water availability is limited, and farmers need to use this resource rationally.

- Inefficient use of water: Traditional irrigation methods can waste water, reducing the efficiency with which this resource is used.

- Degradation of water quality: Pollution and the inappropriate use of fertilisers and pesticides can contaminate water sources, affecting their quality.

Remote Sensing Technologies for Water Resources Management

Remote sensing offers various tools and technologies that can help in the efficient management of water resources. Let's explore some of these technologies and their practical applications in agriculture.

Earth observation satellites

Earth observation satellites are used to monitor the availability and quality of water over large areas. They collect data on soil moisture, evapotranspiration and vegetation cover, which are fundamental for water management.

- Practical example: On a rice farm in Rio Grande do Sul, satellite data is used to monitor soil moisture. With this information, the farmer can adjust irrigation, ensuring that the plants receive the right amount of water without wasting it.

Soil Moisture Sensors

Soil moisture sensors are devices installed in the field that measure the amount of water present in the soil in real time. This data helps farmers determine when and how much to irrigate their plants.

- Practical example: On a coffee plantation in Minas Gerais, soil moisture sensors are used to continuously monitor the amount of water in the soil. Based on this data, the farmer can adjust irrigation precisely, avoiding both excess and lack of water.

Drones Equipped with Sensors

Drones equipped with imaging and infrared sensors can fly over crops and collect detailed data on plant health and soil moisture. They are especially useful for monitoring specific areas and identifying localised problems.

- Practical example: On a fruit plantation in the interior of São Paulo, drones are used to monitor soil moisture in different parts of the field. The data collected helps the farmer identify areas that need more or less water, optimising the use of water resources.

Hydrological Models

Hydrological models use remote sensing data and other environmental information to simulate the water cycle in a specific area. These models help farmers predict water availability and plan irrigation efficiently.

- Practical example: On a sugar cane farm in the north-east, a hydrological model is used to predict water availability during the dry season. Based on the forecasts, the farmer can adjust the irrigation schedule, ensuring that the plants receive enough water during periods of scarcity.

Benefits of Remote Sensing in Water Resources Management

The use of remote sensing in water resource management offers several benefits that contribute to the efficiency and sustainability of agricultural operations.

Efficient Use of Water

Remote sensing technology allows farmers to use water more efficiently, reducing waste and improving crop yields.

- Practical example: On a vegetable plantation in southern Brazil, the use of soil moisture sensors enabled precise irrigation. With the proper application of water, the plants grew healthily and productivity increased significantly.

Cost reduction

Efficient water use can also result in cost savings, as farmers save on water application and energy for pumping and distributing the water.

- Practical example: On a wheat farm in Rio Grande do Sul, the adoption of hydrological models and soil moisture sensors resulted in a significant reduction in water consumption. This saving of resources has translated into a reduction in operating costs, increasing the farm's profitability.

Environmental sustainability

The efficient management of water resources contributes to environmental sustainability, helping to preserve water sources and reduce environmental degradation.

- Practical example: On a grape plantation in the São Francisco Valley, the use of remote sensing technologies has made it possible to optimise water use and reduce environmental impact. With management practices more sustainable, the winegrower has contributed to the preservation of the region's water resources.

Improving Water Quality

Continuous monitoring of water quality through remote sensing technologies allows for the early identification of pollution problems and the implementation of corrective measures.

- Practical example: On a rice plantation in Rio Grande do Sul, the use of Earth observation satellites made it possible to monitor the quality of the water used for irrigation. By identifying the presence of pollutants, the farmer was able to adopt measures to improve water quality, guaranteeing plant health and food safety.

Challenges and considerations

Although remote sensing offers many benefits for water resource management, there are also challenges and considerations to take into account.

Access and Cost of Technology

Acquiring and implementing remote sensing technologies can represent a significant cost, especially for small farmers. However, access to these technologies is becoming more affordable, with financing options and support programmes available.

- Practical example: On a maize plantation in Paraná, the farmer took part in a funding programme to purchase soil moisture sensors. With the financial support, he was able to implement the technology and improve water management on his farm.

Data Interpretation

The effective use of remote sensing technologies requires the ability to interpret the data collected and make informed decisions based on it. Training and capacity building for farmers is essential to maximise the benefits of these technologies.

- Practical example: On a soya farm in Mato Grosso, a farmer took part in a training course on the use of drones and soil moisture sensors. With the knowledge he acquired, he was able to interpret the data effectively and adjust his management practices to optimise water use.

Integration with Existing Management Practices

Integrating remote sensing technologies with existing management practices can be challenging. It is important that farmers adapt their management practices to make the most of the data and insights provided by remote sensing technologies.

- Practical example: On a fruit plantation in southern Brazil, the farmer adjusted his irrigation practices based on data collected by soil moisture sensors and drones. This integration resulted in more efficient water management, improving plant health and productivity.

Climate Variability

Climate variability can affect water availability and the effectiveness of water management practices. It is important that farmers consider this variability when planning and implementing their water management strategies.

- Practical example: On a coffee plantation in Minas Gerais, the farmer adjusted his water management plan based on climate forecasts and remote sensing data. With this information, he was able to adapt his irrigation practices to deal with climate variability and ensure plant health.

Conclusion

The efficient management of water resources is fundamental to the productivity and sustainability of agriculture. Remote sensing offers a series of tools and technologies that make it possible to monitor and manage water accurately and in real time.

In this chapter, we explore how remote sensing can be used in water management, highlighting practical examples and tangible benefits. By adopting these technologies, farmers can increase water use efficiency, reduce costs, operate more sustainably and make informed decisions based on data.

Although there are challenges in implementing these technologies, the long-term benefits often outweigh the initial costs and obstacles. With remote sensing, agriculture can take a significant step towards a more efficient, productive future

CHAPTER 13

REMOTE SENSING APPLICATIONS IN PEST AND DISEASE MANAGEMENT

Hello again! In the previous chapter, we explored how remote sensing can be used in water management. Now, let's look at another crucial topic in agriculture: pest and disease management. Effective control of these factors is essential to ensure plant health and maximise crop yields.

The Importance of Pest and Disease Management

Pests and diseases are a constant threat to agriculture, causing significant losses in productivity and affecting product quality. Effective management of these problems not only protects crops, but also contributes to environmental sustainability and food security.

Pest and Disease Management Challenges

Pest and disease management involves a number of challenges:

- Early identification: Detecting pests and diseases early on is crucial to preventing their spread and minimising damage.

- Effective treatments: Selecting and applying appropriate treatments requires knowledge and precision.

- Resistance: Some pests and diseases can develop resistance to pesticides, requiring integrated and innovative management strategies.

Remote Sensing Technologies in Pest and Disease Management

Remote sensing offers a number of tools and technologies that can help farmers detect, monitor and manage pests and diseases more efficiently. Let's explore some of these technologies and their practical applications in agriculture.

Satellite Images

Satellite images provide a broad and detailed view of crops, allowing early detection of areas affected by pests and diseases. These images capture data in various spectral bands, revealing signals that are not visible to the naked eye.

- Practical example: In a soya plantation in Mato Grosso, satellite images are used to identify areas with water or nutritional stress, which could be signs of pest or disease infestations. With this information, the farmer can target field inspections and specific treatments.

Drones Equipped with Multispectral Cameras

Drones equipped with multispectral cameras can fly over crops and capture detailed images in different spectral bands. These images help identify areas affected by pests and diseases before the symptoms are visible to the naked eye.

- Practical example: On a grape plantation in the São Francisco Valley, drones are used to monitor the health of the plants. The images captured reveal signs of pest infestations, allowing the grape grower to act quickly and apply specific treatments to the affected areas.

Field Sensors

Field sensors, such as chlorophyll sensors and plant health sensors, can detect changes in plant physiology caused by pests and diseases. These sensors provide data in real time, allowing for a quick and accurate response.

- Practical example: In a maize plantation in Paraná, chlorophyll sensors are used to monitor plant health. When the sensors detect a decrease in chlorophyll levels, the farmer is alerted to the possible presence of pests or diseases, allowing immediate action to be taken.

Forecasting Models

Forecasting models use remote sensing data and other environmental information to predict the emergence and spread of pests and diseases. These models help farmers plan and implement preventative management strategies.

- Practical example: On a wheat plantation in Rio Grande do Sul, a forecasting model is used to predict outbreaks of wheat rust. Based on the forecasts, the farmer can adjust the fungicide application schedule, protecting the plants more effectively.

Benefits of Remote Sensing in Pest and Disease Management

The use of remote sensing in pest and disease management offers several benefits that contribute to the efficiency and sustainability of agricultural operations.

Early Detection

Remote sensing technology enables early detection of pests and diseases, before the symptoms are visible to the naked eye. This enables a rapid response, minimising damage and preventing spread.

- Practical example: In a fruit plantation in the interior of São Paulo, drones equipped with multispectral cameras detected early signs of mite infestations. With this early detection, the farmer was able to apply specific treatments and prevent the pest from spreading.

Reduced Use of Pesticides

With precise, localised detection of pests and diseases, farmers can apply treatments only to the affected areas, reducing the use of pesticides and minimising environmental impact.

- Practical example: On a vegetable plantation in southern Brazil, field sensors and satellite images were used to identify specific areas with aphid infestations. With this targeted approach, the farmer was able to significantly reduce the amount of pesticides used.

Better decision making

Remote sensing provides detailed and accurate data that helps farmers make informed decisions about pest and disease management. With this information, they can plan and implement more effective management strategies.

- Practical example: On a soya plantation in Mato Grosso, the use of forecasting models and satellite images enabled the farmer to plan the application of insecticides based on predictions of caterpillar outbreaks. This approach has resulted in more effective plant protection and a reduction in the use of chemicals.

Environmental sustainability

Efficient pest and disease management using remote sensing contributes to environmental sustainability by reducing the impact of pesticides and promoting more responsible agricultural practices.

- Practical example: On a grape plantation in the São Francisco Valley, the use of drones and field sensors enabled a more precise and sustainable approach to pest control. With the targeted application of treatments, the grape grower reduced the environmental impact and promoted the health of the local ecosystem.

Challenges and considerations

Although remote sensing offers many benefits for pest and disease management, there are also challenges and considerations to take into account.

Access and Cost of Technology

Acquiring and implementing remote sensing technologies can represent a significant cost, especially for small farmers. However, access to these technologies is becoming more affordable, with financing options and support programmes available.

- Practical example: On a maize plantation in Paraná, the farmer took part in a financing programme to buy drones and field sensors. With financial support, he was able to implement the technology and improve pest and disease management on his farm.

Data Interpretation

The effective use of remote sensing technologies requires the ability to interpret the data collected and make informed decisions based on it. Training and capacity building for farmers is essential to maximise the benefits of these technologies.

- Practical example: On a soya farm in Mato Grosso, a farmer took part in a training course on the use of drones and forecasting models. With the knowledge he acquired, he was able to interpret the data effectively and adjust his management practices to optimise pest and disease control.

Integration with Existing Management Practices

Integrating remote sensing technologies with existing management practices can be challenging. It is important that farmers adapt their management practices to make the most of the data and insights provided by remote sensing technologies.

- Practical example: On a fruit plantation in southern Brazil, the farmer adjusted his pest control practices based on data collected by drones and field sensors. This integration resulted in more efficient pest management, improving plant health and productivity.

Climate Variability

Climate variability can affect the effectiveness of pest and disease management practices. It is important that farmers consider this variability when planning and implementing their management strategies.

- Practical example: On a coffee plantation in Minas Gerais, the farmer adjusted his pest management plan based on climate forecasts and remote sensing data. With this information, he was able to adapt his practices to deal with climate variability and guarantee the health of the plants.

Conclusion

Effective pest and disease management is fundamental to agricultural productivity and sustainability. Remote sensing offers a series of tools and technologies that make it possible to detect, monitor and manage pests and diseases accurately and in real time.
In this chapter, we explore how remote sensing can be used in pest and disease management, highlighting practical examples and tangible benefits. By adopting these technologies, farmers can improve the efficiency of pest and disease control, reduce costs, operate more sustainably and make informed decisions based on data.

Although there are challenges in implementing these technologies, the long-term benefits often outweigh the initial costs and obstacles. With remote sensing, agriculture can take a significant step towards a more efficient, productive and sustainable future.

CHAPTER 14
THE ROLE OF REMOTE SENSING IN PRECISION AGRICULTURE

Hello, dear reader! We've reached a crucial point in our journey through remote sensing in agriculture. In this chapter, we'll explore how this technology integrates seamlessly with precision agriculture, an approach that promises to radically transform the way we grow our food. Precision farming is not just a trend; it's a silent revolution that is shaping the future of agricultural production.

The Concept of Precision Farming

Before we delve into the role of remote sensing, it's important to understand what precision agriculture is. It is an approach that uses advanced technology to monitor and manage agricultural variables in a detailed and specific way. Instead of treating a crop as a homogenous unit, precision agriculture recognises and responds to variations within the field, optimising inputs and farming practices.

Main Components of Precision Farming

1. Data collection: Use of sensors, drones, satellites and other technologies to collect data on soil, climate and plant conditions.

2. Data Analysis: Use of software and algorithms to interpret collected data and generate actionable insights.

3. Decision-making: Applying inputs (such as fertilisers, pesticides and water) precisely, based on data and analyses.

Integrating Remote Sensing with Precision Agriculture

Remote sensing is a fundamental piece of the precision agriculture puzzle. It provides the detailed data needed to monitor field conditions in real time and make informed decisions. Let's explore some of the ways in which remote sensing integrates with and enhances precision agriculture.

Plant Health Monitoring

One of the most powerful applications of remote sensing is monitoring plant health. Multispectral and hyperspectral sensors capture data that makes it possible to identify plant stresses, often before visible symptoms appear.

- Practical example: On a coffee plantation in Minas Gerais, drones equipped with multispectral cameras are used to monitor plant health. The data reveals areas with water stress, allowing the farmer to adjust irrigation and prevent further damage to the plants.

Nutrient Management

Remote sensing is also crucial for nutrient management. By analysing images and sensing data, it is possible to identify specific nutritional deficiencies and apply fertilisers precisely, avoiding waste and maximising efficiency.

- Practical example: In a maize field in Mato Grosso, field sensors and satellite images are used to monitor the levels of nitrogen in the soil. Based on this data, the farmer adjusts the fertiliser application, ensuring that the plants receive the right amount of nutrients.

Irrigation Control

Water is a precious resource, and efficient irrigation management is essential for agricultural sustainability. Remote sensing makes it possible to monitor soil moisture and plant water requirements, optimising the use of water.

- Practical example: On a grape plantation in the São Francisco Valley, soil moisture sensors and satellite images are used to monitor the plants' water needs. With this data, the grower adjusts the irrigation precisely, saving water and improving the quality of the grapes.

Pest and Disease Detection

As we saw in the previous chapter, remote sensing is a powerful tool for the early detection of pests and diseases. By integrating this information into precision agriculture, farmers can take preventative measures and apply treatments in a localised way.

- Practical example: In a soya plantation in Rio Grande do Sul, drones and field sensors detect early signs of Asian rust. With this data, the farmer applies fungicides only to the affected areas, reducing the use of chemicals and protecting the crop effectively.

Benefits of Integrating Remote Sensing into Precision Agriculture

The integration of remote sensing in precision agriculture brings a series of benefits that go beyond simply increasing productivity. Let's explore some of these benefits in detail.

Increased Productivity

By providing accurate, real-time data, remote sensing allows farmers to make more informed decisions and optimise their agricultural practices. This results in a significant increase in productivity and efficiency.

- Practical example: On a wheat farm in Paraná, the integration of remote sensing data allowed the farmer to adjust the application of fertilisers and pesticides, resulting in a 20% increase in crop productivity.

Cost reduction

The precision provided by remote sensing helps reduce the use of inputs such as fertilisers, pesticides and water. This not only saves money, but also reduces environmental impact.

- Practical example: On a vegetable plantation in the interior of São Paulo, the use of field sensors and drones to monitor plant health enabled the farmer to reduce the use of pesticides by 30 per cent, saving money and promoting more sustainable agricultural practices.

Environmental sustainability

Precision agriculture, enhanced by remote sensing, contributes to environmental sustainability by minimising the use of inputs and reducing environmental impact. This is especially important in a context of climate change and scarcity of natural resources.

- Practical example: On a rice plantation in Rio Grande do Sul, the use of remote sensing to monitor soil moisture and adjust irrigation resulted in water savings of 25 per cent, contributing to the sustainability of water resources in the region.

Better decision making

The detailed and accurate data provided by remote sensing allows farmers to make informed, evidence-based decisions. This improves the efficiency of farming operations and reduces the risk of errors.

- Practical example: On a soya farm in Mato Grosso, remote sensing data analysis enabled the farmer to adjust crop rotation and management practices, resulting in improved soil health and crop productivity.

Challenges and considerations

Despite the significant benefits, the integration of remote sensing in precision agriculture also presents challenges and considerations that must be taken into account.

Cost and Access to Technology

The cost of acquiring and implementing remote sensing technologies can be high, especially for small farmers. However

However, the trend is for these costs to decrease over time as the technology becomes more accessible and widespread.

- Practical example: On a small coffee farm in Minas Gerais, the farmer took part in a funding programme to purchase drones and field sensors. With the financial support, he was able to implement the technology and improve his farming practices.

Training

The effective use of remote sensing technologies requires specific knowledge and skills. It is essential that farmers receive appropriate training and education to maximise the benefits of these technologies.

- Practical example: On a corn farm in Paraná, the farmer took part in workshops and training courses on the use of sensors and data analysis. With the knowledge he acquired, he was able to interpret the data effectively and adjust his farming practices to improve productivity.

Integration with Existing Management Practices

Integrating new technologies with existing management practices can be challenging. It is important that farmers adapt their practices to make the most of the data and insights provided by remote sensing.

- Practical example: On a fruit plantation in southern Brazil, the farmer adjusted his irrigation and fertilisation practices based on remote sensing data, resulting in a significant improvement in plant health and fruit quality.

Climate Variability

Climate variability can affect the effectiveness of precision farming practices. It is crucial that farmers consider this variability when planning and implementing their management strategies.

- Practical example: On a coffee plantation in Espírito Santo, the farmer adjusted his management plan based on climate forecasts and remote sensing data. With this information, he was able to adapt his practices to deal with climate variability and guarantee the health of the plants.

Conclusion

The future of remote sensing in agriculture is bright, with numerous innovations and emerging technologies that promise to transform the sector. By adopting these technologies, farmers can increase productivity, reduce costs, operate more sustainably and improve food security.

In this chapter, we explore the emerging innovations, the impact of automation and the long-term benefits of remote sensing in agriculture. We also discuss the challenges and future considerations that need to be addressed to maximise the potential of these technologies.

As we move on to the last chapter of this e-book, I hope you are inspired to explore and adopt remote sensing technologies in your own farming practice. The agriculture of the future is full of possibilities, and together we can build a more efficient, sustainable and productive sector.

CHAPTER 15
IMPLEMENTING REMOTE SENSING ON YOUR FARM

Hello, we've reached the final chapter of this e-book. Throughout our journey, we've explored what remote sensing is, its applications, benefits and the future of this technology in agriculture. Now, let's address a crucial aspect: how you can implement remote sensing on your own farm.

Step-by-Step Implementation

Implementing remote sensing may seem complex at first glance, but with proper planning and a clear understanding of the necessary steps, this process can be quite straightforward. Here, I present a step-by-step guide to help you get started.

1. Assess Your Needs and Objectives

The first step is to assess your specific needs and objectives. What do you hope to achieve with remote sensing? Your goals may include increasing productivity, improving resource management, reducing costs, or a combination of these.

- Practical example: On a coffee plantation in Minas Gerais, you may be interested in using remote sensing to monitor plant health and identify areas that need immediate intervention.

2. Choose the right technologies

Based on your objectives, choose the remote sensing technologies that best suit your needs. This can include drones, satellites, ground sensors or a combination of them.

- Practical example: If your farm is located in a large area in Mato Grosso, where access can be difficult, satellites can be a good option for large-scale monitoring. For smaller areas, such as a winery in Rio Grande do Sul, drones can offer greater resolution and flexibility.

3. Investigate Technology Providers

Research different technology providers to find those that offer the products and services that meet your needs. Compare prices, customer support and features offered.

- Practical example: Companies like AgEagle and Parrot offer specialised drones for agriculture, while satellite data providers like Planet Labs and Airbus Defence and Space provide high-resolution images and analysis services.

4. Training and Capacity Building

Make sure you and your team are properly trained to use remote sensing technologies. This can include training in drone operation, interpreting satellite images and using data analysis software.

- Practical example: Agricultural universities and research institutions such as Embrapa often offer courses and workshops on remote sensing technologies and data analysis.

5. Data Collection and Analysis

Once the technologies are in place, start collecting data on a regular basis. Use software tools to analyse this data and gain insights into the health of your crops, soil conditions and other critical factors.

- Practical example: In a soya plantation in Paraná, the use of drones equipped with multispectral cameras can provide data on plant vigour, which is then analysed to identify areas with nutrient deficiencies or pest infestations.

6. Decision-making and Action

Use the insights gained from analysing data to make informed decisions about the management of your farm. This can include adjustments to irrigation, fertiliser application, pest control and other farming practices.

- Practical example: In a fruit plantation in the interior of São Paulo, analysing remote sensing data can reveal that a specific area is suffering from water stress, leading you to adjust the irrigation systems for that area.

7. Continuous Monitoring and Adjustments

Remote sensing is an ongoing process. Keep monitoring your crops and adjusting your farming practices based on the data collected. This will allow you to manage proactively and adapt to changing conditions.
- Practical example: On a wheat farm in Rio Grande do Sul, continuous monitoring with drones can help identify diseases at an early stage, enabling quick and effective interventions.

Benefits of Implementation

The implementation of remote sensing in agriculture offers a series of benefits that can transform the management of your farm and increase its productivity and sustainability.

Better decision making

With accurate, up-to-date data, you can make more informed and effective decisions about the management of your farm.

- Practical example: On a maize farm in Mato Grosso, the use of satellite images to monitor plant health makes it possible to quickly identify problem areas and take corrective action before the problem spreads.

Saving Resources

Remote sensing can help optimise the use of resources such as water, fertilisers and pesticides, resulting in cost savings and less environmental impact.

- Practical example: Automated irrigation systems in a vegetable plantation in southern Brazil can adjust the amount of water applied based on soil moisture data, saving water and improving efficiency.

Increased Productivity

By proactively identifying and treating problems, remote sensing can increase the productivity of your crops, resulting in bigger and healthier harvests.

- Practical example: At a winery in Rio Grande do Sul, the use of drones to monitor the health of the vines can lead to more uniform, high-quality grape production, resulting in better wines and greater profits.

Environmental sustainability

Remote sensing promotes more sustainable agricultural practices, reducing the use of chemical inputs and minimising environmental impact.

- Practical example: The precise application of pesticides and fertilisers on a soya plantation in Paraná, based on remote sensing data, can reduce soil and water pollution, promoting environmental sustainability.

Implementation Challenges

Although the benefits of remote sensing are clear, there are challenges to be overcome in its implementation.

Initial Cost

The initial cost of acquiring and implementing remote sensing technologies can be high, especially for small farmers.

- Solution: Research financing options and subsidies offered by governments and agricultural institutions to help cover initial costs.

Technical Complexity

The use of advanced technologies can be complex and require technical skills that many farmers may not possess.

- Solution: Invest in training and capacity building to ensure that you and your team are prepared to use these technologies effectively.

Integration with Existing Practices

Integrating new technologies with traditional farming practices can be challenging and require adjustments to management methods.

- Solution: Develop a gradual implementation plan, starting with small pilot projects to test and adjust the technologies before expanding them to the entire farm.

Conclusion

Implementing remote sensing on your farm is a significant step towards more modern, efficient and sustainable agriculture. Although there are challenges, the long-term benefits far outweigh the initial difficulties.

In this chapter, we discuss a step-by-step guide to implementing remote sensing, highlighting the technologies available, the training required, data collection and analysis, and data-based decision-making. We also explore the benefits and challenges associated with this implementation.
I believe that by adopting remote sensing, you will be prepared to face the challenges of modern agriculture and take advantage of the opportunities that this technology offers. Agriculture is constantly evolving, and remote sensing is a powerful tool to help you remain competitive and sustainable in the future.

Thank you for following this e-book so far. I hope the information presented has been useful and inspiring, and I wish you every success in implementing remote sensing on your farm. Let's build a more efficient, productive and sustainable agriculture together!

REFERENCES BIBLIOGRAPHIC

Ahmed, Asif, Shaik, Nazeer, Shaik, C., Geetha, Priya (2024). 1. Predictive Modelling in Remote Sensing Using Machine Learning Algorithms. International Journal of Current Science Research and Review, doi: 10.47191/ijcsrr/v7-i6-62

Claudia, Aviles, Toledo, Melba, M., Crawford, Mitchell, R., Tuinstra (2024). 1. Integrating multi-modal remote sensing, deep learning, and attention mechanisms for yield prediction in plant breeding experiments. Frontiers in Plant Science, doi: 10.3389/fpls.2024.1408047

D., Anuradha., Ramu, Kuchipudi., B, Ashreetha., Janjhyam, Venkata, Naga, Ramesh., Rami, Ayadi. (2023). 8. Enhancing Agricultural Yield Forecasting with Deep Convolutional Generative Adversarial Networks and Satellite Data. International Journal of Advanced Computer Science and Applications, doi: 10.14569/ijacsa.2024.0150269

Florian, Teste, Hugo, Gangloff, Philippe, Ciais, David, Malowski (2024). 6. Improving early crop yield and price predictions using satellite imagery with machine and deep learning techniques. doi: 10.5194/egusphere-egu24-2028

Hiba, Najjar., Miro, Miranda., Marlon, Nuske., Ribana, Roscher., Andreas, Dengel. (2024). 5. Explainability of Sub-Field Level Crop Yield Prediction using Remote Sensing. doi: 10.48550/arxiv.2407.08274

J., S., Kumar., P., S., Santhosh., T., Kowshik., G., Maheswari. (2024). 7. Deep Learning Based Crop Yield Prediction and Disease Identification. doi: 10.1109/adics58448.2024.10533466

Jonghan, Ko., Taehwan, Shin., Jiwoo, Kang., Jaekyeong, Baek., Wan-Gyu, Sang. (2024). 2. Combining machine learning and remote sensing-integrated crop modelling for rice and soybean crop simulation. Frontiers in Plant Science, doi: 10.3389/fpls.2024.1320969

Khadija, Meghraoui, Imane, Sebari, Juergen, Pilz, Kenza, Ait, El, kadi, Saloua, Bensiali (2024). 9. 9. Applied Deep Learning-Based Crop Yield Prediction: A Systematic Analysis of Current Developments and Potential Challenges. Technologies (Basel), doi: 10.3390/technologies12040043

Khilola, Amankulova, Nizom, Farmonov, Khasan, Omonov, Mokhigul, Abdurakhimova, László, Mucsi (2023). 7. Integrating the Sentinel-1, Sentinel-2 and Topographic data into soybean yield modelling using Machine Learning. Advances in Space Research, doi: 10.1016/j.asr.2024.01.040

Mahlatse, Kganyago., Clement, Adjorlolo., Paidamwoyo, Mhangara., L., Tsoeleng. (2024). 6. Optical remote sensing of crop biophysical and biochemical parameters: An overview of advances in sensor technologies and machine learning algorithms for precision agriculture. Computers and Electronics in Agriculture, doi: 10.1016/j.compag.2024.108730

Mamta, Kumari., Suman, Suman., D., Prasad. (2024). 3. Crop Yield Prediction using Remote Sensing: A Review.doi: 10.1109/iccica60014.2024.10584591

Ning, Wang, Zhong, Ma, Pengcheng, Huo, Xi, Liu, Zhao, He, Kedi, Lu (2023). 10. 3D Convolutional Neural Network with Dimension Reduction and Metric Learning for Crop Yield Prediction Based on Remote Sensing Data. Applied Sciences, doi: 10.3390/app132413305

P, Seralathan, Shirly, Edward (2024). 9. Revolutionising Agriculture: Deep Learning-Based Crop Monitoring Using UAV Imagery - A Review. 9. Revolutionising Agriculture: Deep Learning-Based Crop Monitoring Using UAV Imagery - A Review. doi: 10.1109/icoeca62351.2024.00139

Pradosh, Kumar, Parida., E., Somasundaram., Ramanujam, Krishnan., Suraj, Radhamani., Sivakumar, Uthandi., E., Parameswari., R., Raja., S.R., Shri, Rangasami., Sundapalayam, Palanisamy, Sangeetha., R., Gangai, Selvi. (2024). 4. Unmanned Aerial Vehicle-Measured Multispectral Vegetation Indices for Predicting LAI, SPAD Chlorophyll, and Yield of Maize. Agriculture, doi: 10.3390/agriculture14071110

Priyanka, K., D., R., Umesh. (2024). 4. Deep Learning Approaches for Enhanced Precision in Crop Yield Prediction. doi: 10.1109/incet61516.2024.10593483

R., Priya., U., Rahamathunnisa. (2024). 2. A Comprehensive Study of Remote Sensing Technology for Agriculture Crop Monitoring. Nature Environment and Pollution Technology, doi: 10.46488/nept.2024.v23i02.035

Rabi, N., Sahoo., R., G., Rejith., Shalini, Gakhar., Jochem, Verrelst., Rajeev, Ranjan., Tarun, Kondraju., M., C., Meena., Joydeep, Mukherjee., Anchal, Dass., Mahesh, Kumar., Raju, Dhandapani., Viswanathan, Chinnusamy. (2024). 5. Estimation of wheat biophysical variables through UAV hyperspectral remote sensing using machine learning and radiative transfer models. Computers and Electronics in Agriculture, doi: 10.1016/j.compag.2024.108942

Rahul, Banerjee., Bharti., P., K., Das., Sadaf, Khan. (2023). 8. Crop Yield Prediction Using Artificial Intelligence and Remote Sensing Methods. Advances in geographical and environmental sciences, doi: 10.1007/978- 981-97-0341-8_6

Robin, Young, Rachel, Yuting, Fan (2023). 7. 7. Enhancing Agricultural Productivity Monitoring and Forecasting through Deep Transfer Learning. doi: 10.22541/essoar.169903626.61386533/v1

Rumiana, V., Ray. (2024). 5. A Review on the Current Trends in Crop Yield Forecasting Using Artificial Intelligence. International Journal For Multidisciplinary Research, doi: 10.36948/ijfmr.2024.v06i03.21432

Santanu, Ghosh., Debabrata, Behera., S., Jayakumar., Pulakesh, Das. (2020). 1. Comparison of Sentinel-2 Multispectral Imager (MSI) and Landsat 8 Operational Land Imager (OLI) for Vegetation Monitoring. doi: 10.1007/978-3-030-56542-8_7

Sarabjot, Kaur, Sandhu., S., C., Anand. (2023). 10. Revolutionising Crop Yield Prediction: The Synergy of Remote Sensing and Artificial Intelligence Technologies. Advances in geographical and environmental sciences, doi: 10.1007/978-981-97-0341-8_24

Shahab, Aldin, Shojaeezadeh, Abdelrazek, Elnashar, Tobias, K., D., Weber (2024). 8. Estimating Crop Phenology from Satellite Data using Machine Learning. doi: 10.5194/egusphere-egu24-15347

Shweta, Pokhariyal., N., R., Patel., Ajit, Govind. (2023). 10. Machine Learning-Driven Remote Sensing Applications for Agriculture in India-A Systematic Review. Agronomy, doi: 10.3390/agronomy13092302

MIX
Papier aus verantwortungsvollen Quellen
Paper from responsible sources
FSC® C105338

Printed by Books on Demand GmbH, Norderstedt / Germany